Sustainable Tourism IX

WITPRESS

WIT Press publishes leading books in Science and Technology.
Visit our website for the current list of titles.
www.witpress.com

WITeLibrary

Home of the Transactions of the Wessex Institute.
Papers contained in this volume are archived in the WIT eLibrary in volume 248 of WIT Transactions on Ecology and the Environment (ISSN 1743-3541).
The WIT eLibrary provides the international scientific community with immediate and permanent access to individual papers presented at WIT conferences.
Visit the WIT eLibrary at www.witpress.com.

NINTH INTERNATIONAL CONFERENCE ON
SUSTAINABLE TOURISM

Sustainable Tourism 2020

Conference Chairmen

Juan Casares
University of Santiago de Compostela, Spain
Member of WIT Board of Directors

Francisco Pineda
Complutense University of Madrid, Spain

International Scientific Advisory Committee

Organised by
Wessex Institute, UK

Sponsored by
WIT Transactions on Ecology and the Environment
International Journal of Environmental Impacts
International Journal of Energy Production and Management

WIT Transactions

Wessex Institute
Ashurst Lodge, Ashurst
Southampton SO40 7AA, UK

Editorial Board

Sustainable Tourism IX

Editors

Juan Casares
University of Santiago de Compostela, Spain
Member of WIT Board of Directors

Francisco Pineda
Complutense University of Madrid, Spain

WIT*PRESS* Southampton, Boston

Editors:

Juan Casares
University of Santiago de Compostela, Spain
Member of WIT Board of Directors

Francisco Pineda
Complutense University of Madrid, Spain

Published by

WIT Press
Ashurst Lodge, Ashurst, Southampton, SO40 7AA, UK
Tel: 44 (0) 238 029 3223; Fax: 44 (0) 238 029 2853
E-Mail: witpress@witpress.com
http://www.witpress.com

For USA, Canada and Mexico

Computational Mechanics International Inc
25 Bridge Street, Billerica, MA 01821, USA
Tel: 978 667 5841; Fax: 978 667 7582
E-Mail: infousa@witpress.com
http://www.witpress.com

British Library Cataloguing-in-Publication Data

A Catalogue record for this book is available
from the British Library

ISBN: 978-1-78466-407-7
eISBN: 978-1-78466-408-4
ISSN: 1746-448X (print)
ISSN: 1743-3541 (on-line)

The texts of the papers in this volume were set individually by the authors or under their supervision. Only minor corrections to the text may have been carried out by the publisher.

Sustainable Tourism IX

Editors

Juan Casares
University of Santiago de Compostela, Spain
Member of WIT Board of Directors

Francisco Pineda
Complutense University of Madrid, Spain

WIT*PRESS* **Southampton, Boston**

Editors:

Juan Casares
University of Santiago de Compostela, Spain
Member of WIT Board of Directors

Francisco Pineda
Complutense University of Madrid, Spain

Published by

WIT Press
Ashurst Lodge, Ashurst, Southampton, SO40 7AA, UK
Tel: 44 (0) 238 029 3223; Fax: 44 (0) 238 029 2853
E-Mail: witpress@witpress.com
http://www.witpress.com

For USA, Canada and Mexico

Computational Mechanics International Inc
25 Bridge Street, Billerica, MA 01821, USA
Tel: 978 667 5841; Fax: 978 667 7582
E-Mail: infousa@witpress.com
http://www.witpress.com

British Library Cataloguing-in-Publication Data

A Catalogue record for this book is available
from the British Library

ISBN: 978-1-78466-407-7
eISBN: 978-1-78466-408-4
ISSN: 1746-448X (print)
ISSN: 1743-3541 (on-line)

The texts of the papers in this volume were set individually by the authors or under their supervision. Only minor corrections to the text may have been carried out by the publisher.

Preface

Probably no person committed to the Wessex Institute of Technology, UK, since its first Conference on Sustainable Tourism, held in Segovia (2004), ever thought of circumstances such as those that currently affect this 9th Conference (2020), initially planned to be held in Madrid. We can really think that only very expert epidemiologists and other scientists with solid foundations, and more or less visionary, could foresee a type of pandemic like the current COVID-19. The Planet perhaps has not suffered before, at least as now, an influence of biological origin as serious for the human species as the current plague. The ecological characteristics of this species are closely related to the gravity that the pandemic has reached.

Almost forty years ago, the Spanish ecologist R. Margalef pointed out the following exclusive characteristics of Mankind:

i) Homo sapiens '*consumes a formidable amount of exosomatic energy*' – not food –; this does not occur in any other species;
ii) '*has very effective mechanisms for transporting materials, energy, human populations and other living beings from some places on the Planet to others far away*', both in continental and oceanic environments;
iii) '*this great mobility facilitates the spread of disease and generates pandemics before the disease is only a balanced parasitism – there is a parallel between the evolution of pathogens and the history of the human species*';
iv) '*is capable of seriously contaminating the Earth with waste; among these are new materials not recyclable by natural ecological processes*',
v) *'money has been invented'*, something whose use and consequences seem unparalleled in the functioning of the Ecosphere – the planetary ecosystem. We can add to this two relevant ecological characteristics of human society:
vi) *'build large buildings and cities'* and, curiously,
vii) *'Humanity "says it wants to impose" a brake on its behaviour through the declaration of Protected Natural Spaces'*.

A person interested in the Conferences on Sustainable Tourism can recognize how activities linked to the idea of tourism itself are involved in all the aforementioned characteristics (tourism comprises, according to the World Tourism Organization, "activities – leisure, business and others – of people during their trips and stays in places other than their usual environment for a period of less than one year").

The values recognized by visitors in these places have a relevant aesthetic–cultural component and are the basis of the relationship between visitors and the human societies that created the landscape of the visited territories. The visitor can also recognize the persistence, where appropriate, of remaining landscapes with a wild appearance. However, it is objectively very difficult today to find wild ecosystems that can explain this appearance.

Human societies are part of complex ecological structures in which ecosystems are less and less wild and more cultural. Just two decades ago, ecological science has had metadata and analysis models capable of relating some structures to others and detecting the indicators or 'key factors' of this interdependence relationship.

Tourism currently constitutes a relevant common thread or pathway between these structures and is involved in the interaction between the seven characteristics mentioned above. Money, transportation, energy consumption and its waste are obvious guiding threads in the tourism industry. Now COVID-19 has come into play in a relevant way. The fact that this Conference is virtual, as well as the evident economic, transport and pollution consequences/changes are proof of this.

We believe we know practically the entire Planet thanks to the work that geographers, historians and travelling writers have historically developed, offering images in a graphic and literary way in a very real way. We even imagine very well the Universe itself. Today, especially thanks to cinema and modern media, we can convince ourselves that this is so. But 'Virtual Tourism' is probably still a long way from being tourism reality. We need to travel. We have done it since prehistory with the help of very different means of transportation. We need to continue doing it in our facet as tourists. This same Conference, virtual under the current circumstances, is, in fact, evidence of this.

The Editors
2020

Contents

Contents

LOCAL ACTIVE ENGAGEMENT AS AN EFFECTIVE TOOL FOR SUSTAINABLE TOURISM DEVELOPMENT: FIRST CONSIDERATIONS FROM THE EUROPEAN CULTURAL ROUTES CASE

SABRINA MENEGHELLO & ERICA MINGOTTO
CISET (International Center of Studies on Tourism Economics), Ca' Foscari University, Italy

ABSTRACT

Over the last decade, Cultural Routes (CR) of the Council of Europe (CoE) have flourished as an interesting example of cultural tourism, proposing thematic a mix of natural and cultural heritage and facilitating outdoor ways to experience sustainable tourism practices. The European programme states the importance of partnering among private and public stakeholders in defining and managing the Route as a sustainable tourism product; in fact, most of the studies on CR examine their state of development and future strategies, starting from these partnerships and their ability to reach this goal. Recently, some scholars have been investigating CR capability in creating new models of participation in heritage with shifting interest in different players, i.e. local communities and tourists turning from passive recipients to engaged actors. The paper aims to further investigate the specific role of these actors in CR tourism processes and resulting social impacts. It represents a preliminary investigation originating from the findings of a pilot study commissioned in 2016 by the European Institute of Cultural Routes (EICR), aimed at developing a measurement and quality control tool to ensure sustainability of CR, starting from a group of seven selected CR. In particular, this study investigates those findings, focusing mainly on the contribution of communities and tourists in tourism processes. Additional surveys and semi-structured interviews with CR managers have been conducted to better identify forms of participation in decisions, product design, and tourism practices. Considerations about engagement effectiveness in relation to typology of actors, interactions, and specific phases of the management process, as well as observed social benefits, are among the key findings.
Keywords: sustainable tourism, European cultural routes, local engagement, social benefits.

1 INTRODUCTION

European Cultural Routes (CR) of the Council of Europe (CoE) is a specific transnational programme aimed at promoting tangible and intangible heritage and activating mobility among countries. As a consequence, it aims at playing a primary role in generating positive impacts addressing sustainable development in many tourism destinations [1].

While it is evident that it is bridging a rich material and immaterial heritage with hospitality and welcoming services, the programme needs the presence of local actors and their living culture, what is not yet sufficiently explored is the weight and the effects that their active participation have in territorial processes focusing on tourism. Moreover, among the relevant subjects, tourists have been playing a more and more intense part in local dynamics [2]–[4].

Starting with the description of seven CR of the CoE according to the findings of a pilot study [5] commissioned in 2016 by the European Institute of Cultural Routes (EICR), this paper intends to further investigate the importance of local active engagement as effective tool in CR tourism processes addressing sustainable development, especially with regard to the social components of sustainability affecting people's well-being [6].

WIT Transactions on Ecology and the Environment, Vol 248, © 2020 WIT Press
www.witpress.com, ISSN 1743-3541 (on-line)
doi:10.2495/ST200011

2 CULTURAL ROUTES AND "LANDSCAPE TOURISM"

In the literature, the tourism product based on cultural itineraries is typically analysed as an example of cultural tourism [7], [8]. It is recognized that in defining the nature of CR it has always been relevant the concept of "heritage". However, to better understand the meaning of the European programme for the tourism sector it may be useful to introduce concepts such as "landscape" and "landscape experience" [9]–[14]. They enrich the definition of "heritage" that, in this perspective, could be perceived not only as an expression of human culture but also of human relations with the place in which he functions [15]–[18].

The present study will, therefore, embrace the definition of "cultural landscape tourism" coined by CISET [19] as conceptual framework to analyze tourism planning and management processes of cultural routes. The aim is to focus on the specific CR of the CoE. According to the definition, "cultural landscape tourism" describes a specific expression of cultural tourism, including both the demand side, i.e. the tourist landscapes and the supply side, i.e. the tourism landscapes [18]. It does not define a single tangible or intangible heritage element but rather the whole landscape resulting from a mix of environmental, historical, socio-economic characteristics, i.e. a kaleidoscope of cultural past and present meanings and values of a given area. The mentioned mix is variable according to the specific territorial context and identifies the proposed and perceived uniqueness of it. This definition may help in focusing on specific "tourist landscapes" characterized by the tourist search for local authenticities and practices, i.e. enjoying food productions, living minor villages events, choosing slow mobility. The importance of networks, itineraries and hubs – thus qualifying as itinerant tourism rather than sedentary – as well as the centrality of services favouring cultural exchanges with local people – ones who often are not directly part of the tourism system, i.e. farmers, artists, citizens, associations – seem to be among the most evident "tourism landscapes" describing the supply side.

In this perspective the concept of "landscape" better describes the multiple and actual territorial dynamics [20] of which those of the tourism sector are a part [21]. Moreover, this specific concept calls for a careful reflection on the role of different stakeholders, their participation [14] and influence in terms of social benefits. Therefore, the focus addresses sustainable tourism development in its social domain.

3 CULTURAL ROUTE, LOCAL ENGAGEMENT AND SOCIAL SUSTAINABLE DEVELOPMENT

Since the 1990s, scholars have been investigating CR of the CoE adopting different perspectives, focusing in particular on the CR as tourism product based on heritage, on the state of development and possible future strategies [22].

CR heritage conservation, tourism product and destination planning and management, networking and branding processes have turned out to be the main key research areas. In particular, most recent publications on CR heritage safeguarding focus on how to implement strategic plan for heritage route protection by the means of information technologies such as G.I.S. [23]. In terms of CR planning and management, specific investigations address to specific strategies, i.e. increasing length of stay [24] or report on visitor motivations and preferences. Concerning the subject of destination branding, researchers have been experimentally focusing on the importance of geo-visualization and texts as strategic tools to explore how the routes are perceived by tourists [25].

In the end, state of the art of the most recent contributions on this European programme leads to the conclusion that while much of this work has provided important practical management insights, it has tended to do so in ways that maintain or facilitate the legitimacy of the top-down management approach to heritage places.

The present study is based on the awareness that recently some scholars have been in parallel investigating on the CR capability to create new models of participation, in a bottom up approach, turning local communities and tourists from passive recipients to engaged actors. From the idea that the Route is first of all an area of cultural, social and civic activities and that the investigation should focus on how Routes are understood in local communities, the work of Anna Goral [15] is remarkable for the capability to clearly address new issues. Since 2016 Goral have been moving behind the concept of Route as mix of resources to focus on the role of people, by defining CR "not only tourism products, but structures which affect and are affected by local communities" [20]. Moreover, she points out the importance of tourists together with local communities in these processes.

Moreover, literature has paid less attention, in terms of both quantitative and qualitative research, on how local empowerment of communities and tourists can be an effective tool in CR sustainable tourism development addressing social issues.

Social components of sustainability can be observed in specific areas of intervention able to impact community well-being. Some scholars define these components addressing people well-being through the following categories: "people" or "connectedness", i.e. social support, social trust, civic engagement; "place" or "liveability", i.e. emotional attachments such as a sense of belonging, cultural heritage, aesthetics and also access to job, education, housing, transport, leisure, public safety; and "power" or "equity", i.e. political voice, inclusion, diversity, equal access to local resources and opportunities [26].

The most recent tourism management literature explains why is important to focus on this specific theme and how the substantive (ends) and procedural (means) aspects defining social sustainability intertwine [27]. As far as the category of "livability" concerns, some studies describe that mediating engagement with local culture during tourism processes influences how authentic the experience is both for local people and tourists [28]; with reference to the categories of "connectedness" and "equity", some researchers underline that planning and stimulating different interactions among actors positively affects the design of significant experiences [29], bringing emotional benefits for residents [30] and tourists [3], [31], [32].

This preliminary study focuses on the weight of local active engagement as effective tool in CR tourism development addressing social sustainability. In more detail, investigation aims at understanding whether and in which stage local participation play a role in tourism planning and management processes.

The research presentation has been divided in two paragraphs. The first part (paragraph 4) discusses the main findings of a EICR pilot study for assessing sustainable development along seven selected CR of the CoE. The second part (paragraph 5) aims at demonstrating the strategic role of local engagement in defining CR not only as a tourism product but also as place of people interactions and social exchanges, i.e. a "cultural landscape laboratory".

4 THE EICR PILOT STUDY

4.1 Reasons, steps and methodological approach of the pilot study

The Council of Europe, the 70-year-old international organisation whose stated aim is to uphold human rights, democracy and the rule of law in Europe, began to promote and develop the programme devoted to Pan European Cultural Routes, during the 1980s, by establishing in particular three main objectives: (1) making European citizens aware that there is an European cultural identity; (2) preserving and enhancing the European cultural heritage; (3) supporting the development and enhancement of cultural tourism in Europe [33]–[35]. While cultural routes in Europe have long existed before the institution of the CoE programme, they have been recognized as key strength also for the development of local, regional and transnational tourism only in the last three decades [1], [36].

Since the launch of the Programme in 1987 with the Declaration of Santiago de Compostela, the Council of Europe described it "as an invitation to travel and to discover the rich and diverse heritage of Europe by bringing people and places together in networks of shared history and heritage". Therefore, EICR launched in 2014 a feasibility study for a system of indicators aimed at assessing CR. Five fields of actions/indicators have been selected: (1) Governance (how the partners manage the network); (2) Communication (exchange between the CR members and with the public); (3) Sustainable management; (4) Indicators of cultural action; (5) Economic impact. Great importance was paid also to tourism sustainable development along the CR, in order to ensure the economic, socio-cultural and environmental sustainability. Then, the Council of Europe co-operated with the European Commission in 2013 within the field of the development of ETIS – the European Tourism Indicator Systems promoted by the European Commission, with the scientific support of a pool of experts. ETIS, whose final version was launched in 2016, is both a management tool, that any destination can implement on a voluntary basis to take a sustainable approach to destination management, and a monitoring system, easy to use for collecting data and detailed information.

In the framework of this co-operation, some of the experts who participated in the ETIS task force, were involved in 2016 by EICR in the pilot project to adapt ETIS indicators for use by the CR, taking into account their characteristics and specificities [5]. At the invitation of the task force coordinator, one of the authors, was involved in the study and had the opportunity to participate in the implementation of some activities and in the analysis of findings. The methodological approach proposed by the pool of experts for adapting ETIS to CR and measure the economic, socio-cultural and environmental sustainability of the CR was based on a horizontal approach (governance model), related to the organizations responsible for the management of the Routes.

The main steps of the pilot study that took place from May to December 2016, can then be summarised as follows.

1. Adaptation of the ETIS indicators to the specificities of CR and development of a specific tool, i.e. the Governance Model Questionnaires addressed to CR managers.
2. Selection and involvement in the pilot study of seven CR for testing the tools.
3. Distribution of the Governance Model questionnaire among the organizations managing the seven CR selected.
4. Collection of questionnaires and analysis of results.

The CR involved were: (1) Transromanica – The Romanesque Routes of European Heritage; (2) Iter Vitis – Les Chemins de la Vigne en Europe; (3) Via Francigena; (4) Saint James Way – Santiago de Compostela; (5) European Routes of Jewish Heritage; (6) Route of Olive Tree; (7) Route of St. Olav Ways.

The Governance Model Questionnaire was developed in order to assess the contribution of the governance model of each CR to the economic, socio-cultural and environmental impacts in the territories along the Route. It was based around seven main themes: (1) Legal status; (2) Territorial dimension; (3) Membership typology; (4) Members communication and cooperation; (5) Main financial sources; (6) Local community involvement; (7) Main activities and specific projects in which funds are invested, in particular in terms of cultural heritage promotion, enhancement of people awareness, tourism development, promotion of sustainable and responsible tourism, social and economic development.

4.2 Main findings coming from the pilot study

Some findings coming from this pilot study [5] have provided useful inputs for the specific scope of this paper. For that reason, the main results of this first analyses are presented hereafter.

1. Legal status: five CR are non-profit associations (Transromanica, Iter Vitis, Via Francigena, Saint James Way, European Routes of Jewish Heritage), one CR is a foundation (Route of Olive Tree) and one is a government organisation (Route of St. Olav Ways).
2. Territorial dimension: due to the transnational dimension all CR involve from three to more than 20 countries and from 10 to 100 main destinations.
3. Membership: in addition to public entities, all CR includes different typologies of bodies, i.e. non-profit organisations and/or private companies and/or foundations. The only exception is Saint James Way that involves public authorities only.
4. Members communication and cooperation: all CR ensure the communication and exchange among members, through meetings throughout the year (from two to six meetings) and/or a general assembly. Regarding cooperation, different actions are implemented: four CR promote joint events to stimulate the evolution of the network; four CR support destination strategies through tourism product development; six CR develop promotional and training activities.
5. Main financial sources: all CR rely on diversified sources of income, based in general on members' fee and/or EU, national and regional co-funding, and/or self-financing (merchandising, sales of good and services), and/or sponsorships and donations. In particular, five CR use EU co-funding and three use also self-financing.
6. Local community involvement: all CR directly organise or support its members in organising activities in favour of local involvement, mainly cultural events and public conferences.
7. Main activities and specific projects: CR use their funds for organising or supporting members to organise events and for investments, such as: existing buildings renovation/reuse (Iter Vitis); support to destinations and companies who put in place certification/labelling schemes (Iter Vitis and Jewish Heritage); promotion of local attractions and development of local tourism industry and/or local labour market (Transromanica, Francigena and Olive Tree); development of other economic sectors linked to tourism (St. Olav).

Another finding coming from the pilot is the identification of the following hypotheses that are directly or indirectly connected with the involvement of local communities and tourists in the CR development processes.

1. The involvement and engagement of the local community are strategic for supporting a participative process which takes into account local specificities and needs; making local residents, companies, public institutions, etc. aware of living in an area that is part of a CR; for stimulating social integration. The more active the CR organisation is in involving the local community, the greater the social benefits could be.
2. A strong cooperation among partners and with other local bodies, could support the exchange of good practices, the sharing of common strategies and actions in terms of socio-cultural growth, development of other networks and partnerships outside the CR. The level of cooperation among the members of the organisation could be considered as an input for generating social benefits.

Due to these results it has seemed to be important to go further to a second step of the analyses and investigate more in detail the question of local engagement.

5 THE STUDY ON LOCAL ENGAGEMENT IN CULTURAL ROUTES

5.1 Methodology

Based on a multidimensional approach [37] not limited to the concepts of "stakeholders" and "networking" derived from management studies [38], [39], a questionnaire has been created and sent to all 38 CR managing organisations [40]. The seven cases analysed in the previous research have been contacted first in order to continue the investigation with them. Since three out of seven have not replied within the specific deadline and due to the coincidence of the data collection with the beginning of the pandemic emergency, the authors have not considered appropriate to insist with the missing respondents. Therefore, they have decided to consider the whole list of CR with the aim of collecting information from the selection of those available to participate in the second part of the study.

The main steps of the study that took place from November 2019 to the beginning of March 2020, can then be described as follows:

1. Definition of the structure of a questionnaire aiming at collecting data and information about the role of local engagement in tourism processes.
2. Creation of the online version of the questionnaire [41].
3. Collection of contact information of all the 38 CR from the European Cultural Route Official Web Site.
4. Distribution of the questionnaire among the managing organizations.
5. Collection of seven online filled questionnaires.
6. Analysis of results.

The questionnaire has been structured in six sections consisting of closed, multiple and open questions aimed at collecting data and considerations from CR managers about the following aspects: (a) Information on the CR from a tourist point of view, i.e. level of development, typical tourist experience, offer and demand, visitor profile; (b) Weight of specific categories of actors, i.e. local communities, tourism and cultural professionals, and operators from other economic sectors, in tourism processes. In this session it was asked to select for each category the three most relevant subjects in involvement; (c) Engagement in different phases of the process. The phases have been divided into three successive steps, each comprising different actions. The first step concerns the definition of the project objectives and strategies to achieve them (the decision-making sphere). The second phase concerns the definition and testing of the tourism experience, the so-called tourist product (the creative sphere). The third phase concerns the actual realization of the tourism experience and the practice by different audiences (the implementation sphere) with the real meeting and cultural exchange between "insiders" and "outsiders". Tourists/visitors have also been considered to assess their role throughout the process; d) Role of different types of interactions per phase: operators vs local communities (b2b), operators vs visitors/tourists (b2c), tourists vs tourists (c2c). Importance and efficiency of each specific interaction has been assessed in the final sections.

These are the respondent CR: (1) Via Francigena (the Italian case), (2) Via Francigena (the French case), (3) the Santiago de Compostela Pilgrim Routes (4) Iter Vitis, (5) the Phoenician Route, (6) the Route of St. Olav Ways, (7) the European Route of Megalithic Culture. This study has to be intended as an exploratory investigation.

5.2 Main findings coming from the local engagement analyses

5.2.1 Tourism indications

The general tourism considerations collected from the seven CRs helps to contextualize the results obtained about the specific focus.

According to survey, the role of communities and tourists as active players does not emerge as a first answer but after having gone into the matter with more detailed questions. At first sight the most relevant actors indicated for the definition of tourism experiences are the private small and medium enterprises (SMEs) along the Route, generally family-based businesses including accommodation owners. In the following paragraphs it will be highlighted how the answers and the results obtained about the active role of other subjects are more articulated.

Concerning the tourist offer defining the Route it is not possible, in their opinion, to define specific sites as the most attractive ones. As stated by the Director of the Phoenicians' Project "the Route is a journey without goals, where the single attractor is not important, what is important is the sense of the Route in its complexity and atmosphere. We therefore consider the entire Route as an attraction". Some of the respondents use the term "landscape" to describe this complexity.

With regard to the characteristics of the demand, the CR generally activate a greater number of international tourists, about 60% vs 40% of domestic flows. The experience of the CR satisfies various motivations of "mature" travelers. "In the Via Francigena" explains the General Manager "spirituality, knowledge of the self and intimate research emerges. The religious reason is fixed at the 15%". The Director of the Phoenicians' Route adds that the Route "is oriented towards experiential and creative motivation, in which the guest is involved". Other expressions used by respondents to describe motivations, tourists profile and experiences can be associated to the following four conceptual categories: (a) "time" meaning memory, "distant past" experiences, discovery of heritage related to history; (b) "space and its elements" meaning "silent" nature, legends and sagas, well-being; (c) "mobility and its expressions" referring to hiking, walking, trekking, bike, horse, slow tourism experiences; (d) "a combination of natural and anthropic attributes of the contexts", i.e. landscape, knowledge of the cultures.

5.2.2 Different categories of actors

Five categories of actors, i.e. stakeholders can be defined: public institutions, tourism businesses, other sectors, host communities and tourists [42]. They are groups or individuals who are associated with tourism development initiatives and therefore can affect or be affected by the decisions and activities concerning those initiatives.

The second session investigates who are different actors in the whole process.

- With regard to host communities, it emerged the relevance of cultural associations and, secondly, citizens who have interest in developing a tourism business. Citizens with focused interests (social, environmental) and related associations are quite relevant;
- Considering public and private operators, institutions are the most quoted. Accommodation sector is assessed as relevant whereas cultural private management staff and information offices are quite relevant. Only one respondent quotes as relevant the incoming travel agencies. No citation for the transport operators;
- As far as the role of other actors not directly operating in the tourism sector, farmers are the most quoted. Artisans are considered relevant. Other mentioned subjects are: schools, universities, chambers of commerce, public and private foundations, media (press,

social, televisions, radio), nautical charters, port authorities, pilgrim associations. They are more related to the territorial specificity of single projects;
– From this first part of the survey tourists are not perceived and mentioned as local actors.

5.2.3 Weight of actors versus phases of the process
Concerning the engagement of the different actors in the specific phases of the process the following emerges (Fig. 1):

– the first phase of the decision-making process highlights the primary role of two types of actors. On the one hand, public institutions have a primary role in heritage safeguarding alongside the cooperation of volunteers from local communities and in organizing meetings aimed at sharing decisions supported by private operators and local communities. On the other hand, professional tourism operators have a primary role with regard to both internal and external marketing activities often with the organizational and financial support of public institutions. Private sector has a key role also in carrying out surveys and collection of data and opinions about specific tourism aspects of the CR. Locals and tourists are cited as survey respondents and therefore considered "engaged" subjects. Local communities are the most active "players" in giving voice to specific subjects, i.e. disadvantaged people facing disabilities and/or risking social exclusion asking for job positions, and needs i.e. instances to avoid land degradation, heritage banalization, social marginalization;
– as far as the definition and testing of tourism experiences is concerned, public institutions are considered the main subject in sharing heritage contents, communicating local values, organizing meetings and training sections for planning or re-assessing new proposals. The role of the public institutions is also important in indicating opportune experiences, useful data on tourist motivations and profiles. Here the weight of the local communities as supporters emerges as relevant. Their engagement concerns the indication of selected contents, places, itineraries and suppliers to be privileged for dedicated services and messages and the direct test of the experiences conceived. Finally, tourism private operators take part in planning meetings and training courses, in structuring proposals with regard to the technically confirmation of suppliers, places, itineraries and in reporting on customer preferences. Their weight is more evident in the actions of choice and sharing, in concert with public institutions, of promotional messages through the various channels;
– the third step concerning the concrete realization of the experience and the practice by different audiences sees the most relevant role for local communities. Citizens and associations act in an evident way in explaining heritage elements, telling stories and legends, providing information, animating experiences such as singing, playing, dancing, hosting at home, teaching arts and crafts. Important here is the role of tourists as well, especially in their interaction with local communities. Tourists take a decisive part in co-creating specific experiences starting from their cultural background, i.e. enriching religious rituals, culinary practices, making new light to specific heritage aspects suggesting personal interpretation, slowly discovering new landscapes, perceived as important for them. Tourist are also relevant in sharing experiences they are living through videos, photos, descriptions and comments, especially online. Public institutions have a limited role: explaining heritage aspects, welcoming and providing information are managed by authorized figures in charge, i.e. local guides. Even the role of private operators is mentioned limited to accommodation and assistance services,

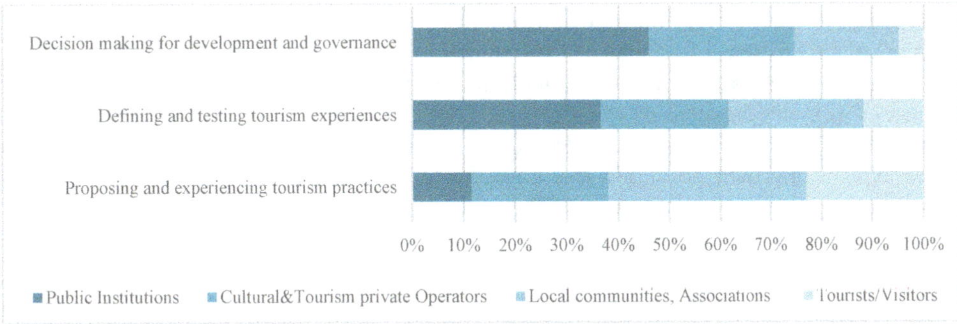

Figure 1: Weight of involvement of different actors in the CR tourism definition and management process.

technical indications and solutions, and promoting contents also through videos and web channels.

As far as the analysis of the interactions between the different types of actors is concerned, it turns out that the decisive role is played by b2c relations, meaning the dynamics between internal and external subjects. Also the frequency of this specific interaction during the whole process is also evaluated as an important condition for CR success in the tourism field.

The primary role of b2c interactions is particularly evident both in the first decision phase of the process and in the development of tourism experiences. However, in the first phase, when sharing of choices is necessary, also b2b relations are considered relevant. In the third phase of experience implementation the presence of interactions among different actors is considered a less significant condition of process effectiveness. However, c2c interactions are here of interest and are mentioned as significant for the success of the tourist and visitors experience.

6 DISCUSSION OF RESULTS AND FINAL CONSIDERATIONS

The results coming from the first step of analysis raised, among other things, two central hypotheses: the level of cooperation among the members of the CR organisation could be considered as an input for generating social benefits in its territories; the more active the organisation is in involving the local community, the greater the social benefits could be.

Based on these assumptions, the second part of the analysis has investigated how effective local engagement is in CR tourism development and to what extent this engagement generates positive social effects addressing objectives of sustainability.

In the seven CR analysed cases, whereas the concomitant and continuous presence of all actors in the different stage of the tourism development process is evident, the weight of involvement varies in relation to the typology of actors and the phase of the process.

According to the seven interviewed managing organizations, CR tourism development can be effective not only thanks to a variety of actors interacting among them along the whole process but above all thanks to the active involvement of local communities. With regard to their indications, when the projects are managed only by public actors there is a risk of not implementing a proper enhancement of the destinations through tourism development.

For all the analysed cases institutions and tourism enterprises have a leading role in decision making processes (75% weight of involvement). Even if among actors the presence of local communities has been demonstrated to be key, it is a minority in the first decision making phase (20% weight of involvement). The case of the Prehistoric Rock Art Trails is emblematic in this sense. It counts a large number of rural areas lacking an adequate tourism offer that could be based on local farm guest houses if farmers were involved in participatory processes from the first phases.

Also designing and testing tourism products are institution driven processes but here it is more evident that their definition and implementation are only possible through a strong cooperation with local communities (27% weight of involvement).

Finally, local communities seem to be the most important actor in the final part of the process, i.e. in proposing and experiencing tourist practices (about 40% weight of involvement). At this stage locals have a leading role mostly in interacting with tourists who are themselves stimulated to be active actors in the co-creation of tourism experiences.

According to the survey, contribution of local communities generates positive social impacts mostly in the second and third phases. In fact, by supervising suppliers, itineraries and related narratives, citizens and associations help in selecting authentic values from the whole "landscape". Secondly, by testing and realizing tourist experiences, communities can play a relevant role in guaranteeing genuine experiences. Moreover, by creating specific proposals for tourists, local communities implement actions for safeguarding local culture, heritage and environment. As shown by the Phoenician Route case, local communities are able to revitalise crafts and specific expertise, i.e. those related to the fishing sector, as well as the correct interpretation of significant cultural expressions.

Concerning the role of tourists as active subjects, the research has shown that definition and management of CR is assured also by their local active engagement. Alongside the process tourists seems to become indispensable not only because they contribute in promotion with online texts, photos, videos and reviews during the experiences but above all because they are more and more engaged in co-creating parts of the tourist practices and narratives together with local actors. As stated by the Manager of the of Saint Olaf Ways, the project is an example of cultural exchanges and sharing of common values between local actors and tourists. In fact, every year more than 3,000 pilgrims and walkers go to the Saint Olaf Ways across Germany, Sweden or Netherlands and share with local communities a common desire for physical and mental well-being, sense of spirituality and faith and close connection with nature. Strengthening the roles and responsibilities of external visitors in local processes by flavouring the positive feeling of being useful and being able to contribute and take part in local life practices are among social benefits which should be further analysed.

This analysis represents a preliminary and exploratory investigation that could be further studied, first of all trying to engage in the survey also the other CR that were not available to participate in the study at that time. In addition, the set of interviews should be made more comprehensive in order to better understand if differences emerge depending on the type of Route, its level of development and the specific territorial contest. Since the focus of this research is about the engagement and active participation of internal and external subjects, it would be relevant for the analysis to involve themselves in the study, by directly interviewing residents, local operators and associations, visitors and tourists, etc. In such a way there would be the opportunity to directly collect their opinions about how they perceive their role and their involvement in the CR development; to what extent they feel part of the process; what main benefits come them from the CR.

Although implementable, the present results can be useful for researchers, stakeholders and decision-makers who need to plan, guide and monitor processes of local engagement in tourism planning and management aiming at sustainability.

REFERENCES

[1] Murray, M. & Graham, B., Exploring the dialectics of route-based tourism: The Camino de Santiago. *Tourism Management*, **18**(8), pp. 513–524, 1997.

[2] Campos, A.C., Mendes, J., Oom do Valle, P. & Scott, N., Co-creation experiences: attention and memorability. *Journal of Travel & Tourism Marketing*, **33**(9), pp. 1309–1336, 2016.

[3] Kim, J. & Fesenmaier, D.R., Measuring emotions in real time: Implications for tourism experience design. *Journal of Travel Research*, **54**(4), pp. 419–429, 2015.

[4] Pine, J.B. & Gilmore, J.H., *The Experience Economy: Work is Theatre & Every Business A Stage*, Harvard Business School Press: Boston, 1999.

[5] Study report on ETIS implementation within 7 pilot Cultural Routes of the Council of Europe. Cinzia de Marzo (EU legal advisor and senior expert on sustainable tourism and indicators) for the Council of Europe (Confidential report).

[6] Cloutier, S., Ehlenz, M.M. & Afinowich, R., Cultivating Community Wellbeing: Guiding Principles for Research and Practice. *International Journal of Community Well-Being*, **2**, pp. 277–299, 2019.

[7] Csapo, J., The role and importance of cultural tourism in modern tourism industry. *Strategies for Tourism Industry – Micro and Macro Perspectives*, eds M. Kasimoglu & A. Handan, InTechOpen, pp. 201–232, 2012.

[8] Terzić, A., Bjeljac, Ž., Jovičić, A. & Penjišević, I., Cultural route and ecomuseum concepts as a synergy of nature, heritage and community oriented sustainable development. *European Journal of Sustainable Development*, **3**(2), pp. 1–16, 2014.

[9] Berti, E., *Itinerari culturali del Consiglio d'Europa tra ricerca di identità e progetto di paesaggio*, Firenze University Press: Firenze, 2012.

[10] Council of Europe, *European Landscape Convention*, 2000.

[11] Ingold, T., The temporality of the landscape. *World Archaeology*, **25**(2), pp. 152–174, 1993.

[12] Minca, C., The tourist landscape paradox. *Social & Cultural Geography*, **8**(3), pp. 433–453, 2007.

[13] Terkenli, T.S., Landscapes of tourism. *The Wiley-Blackwell Companion to Tourism*, eds A.A. Lew, C.M. Hall & A.M. Williams, 1st ed., John Wiley & Sons: Oxford, pp. 282–293, 2014.

[14] Waterton, E., Whose sense of place? Reconciling archaeological perspectives with community values: Cultural landscapes in England. *International Journal of Heritage Studies*, **11**(4), pp. 309–326, 2005.

[15] Bogacz-Wojtanowska, E. & Góral, A., Networks or structures? Organizing cultural routes around heritage values. *Humanistic Management Journal*, **3**, pp. 253–277, 2018.

[16] Kirshenblatt-Gimblett, B., Intangible heritage as metacultural production. *Museum International*, **56**(1–2), pp. 52–65, 2004.

[17] Smith, L., *The Uses of Heritage*, Routledge: Oxford, 2006.

[18] Terkenli, T.S., Castiglioni, B. & Cisani, M., The challenge of tourism in terraced landscapes. *World Terraced Landscapes: History, Environment, Quality of Life*, M. Varotto, L. Bonardi & P. Tarolli, Environmental History, vol. 9., Springer: Cham, pp. 295–309, 2019.

[19] Meneghello, S. & Mingotto, E., Networks of historic houses as strategic option for sustainable tourism development. The venetian villas case. *International Journal of Sustainable Development and Planning*, **3**, pp. 197–208, 2018.

[20] Góral, A., Cultural routes as a medium for changes in regions. *Sharing Cultures*, ed. S. Lira, R. Amoeda & C. Pinheiro, vol. 2011, Green Lines Institute, pp. 719–728, 2016.
[21] Rabbiosi, C., *Il territorio messo in scena. Turismo, consumi, luoghi*, Mimesis, 2018.
[22] Durusoy, E., *From An Ancient Road to a Cultural Route. Conservation and Management of the Road between Milas and Labraunda*, Institut français d'études anatoliennes, 2014.
[23] Oikonomopoulou, E., Delegou, E.T., Sayas, J. & Moropoulou, A., An innovative approach to the protection of cultural heritage. The case of cultural routes in Chios Island, Greece. *Journal of Archaeological Science: Reports*, **14**, pp. 742–757, 2017.
[24] Rodríguez, X.A., Martínez-Roget, F. & González-Murias, P., Length of stay evidence from the Santiago de Compostela. *Annals of Tourism Research*, **68**, pp. 9–19, 2018.
[25] Jones, E., Guido, D. & Severo, M., Towards place-based exploration of Instagram: Using co-design to develop an interdisciplinary geovisualization prototype. *Journal of Spatial Information Science*, **17**, pp. 1–30, 2018.
[26] Atkinson, S., Bagnall, A., Corcoran, R. & South, J., What is community wellbeing? Conceptual Review, *Technical Report*, 2017.
[27] Helgadóttir, G., Vilborg Einarsdóttir, A., Leah Burns, G., Þóra Gunnarsdóttir, G. & Matthíasdóttir, G.M.E., Social sustainability of tourism in Iceland: A qualitative inquiry. *Scandinavian Journal of Hospitality and Tourism*, **19**(4–5), pp. 404–421, 2019.
[28] Zatori, A., Smith, M.K. & Puczko, L., Experience-involvement, memorability and authenticity: The service provider's effect on tourist experience. *Tourism Management*, **67**, pp. 111–126, 2018.
[29] Ming Ming, S. & Wall, G., Implications of host/guest interactions for tourists-travel behaviour and experiences. *Tourism*, **58**(1), pp. 37–50, 2010.
[30] Presenza, A., Del Chiappa, G. & Sheehan, L., Residents' engagement and local tourism governance in maturing beach destinations. Evidence from an Italian case study. *Journal of Destination Marketing and Management*, **2**(1), pp. 22–30, 2013.
[31] Han, C. & Imram, R., Cultural tourism: An analysis of engagement, cultural contact, memorable tourism experience and destination loyalty. *Tourism Management Perspectives*, **26**, pp. 153–163, 2018.
[32] Millan, C.L., Garzon, D., Navarro, D., C2C interactions creating value in the Route of Santiago. *Journal of Business Research*, **69**(11), pp. 5448–5455, 2016.
[33] *Global Report on Cultural Routes and Itineraries*. Affiliate Members Report: Vol. 12, UNWTO, Madrid, 2015.
[34] Resolution CM/Res 12 on the Cultural Routes of the Council of Europe, 2007.
[35] The ICOMOS Charter on Cultural Routes, 2008.
[36] Council of Europe, *Cultural Routes Management: From Theory to Practice*, 2015.
[37] Ciervo, M., Un approccio geografico per una pianificazione territoriale sostenibile. *Bollettino della Società Geografica Italiana*, **13**, pp. 559–572, 2014.
[38] Birley S., The role of networks in the entrepreneurial process. *Journal of Business Venturing*, **1**, pp. 107–117, 1985.
[39] Freeman, R.E., *Strategic Management: A Stakeholder Approach*, Pitman: Boston, 1984.
[40] Online references on the 38 CR Managing Organizations, Council of Europe. www.coe.int/en/web/cultural-routes/all-cultural-routes. Accessed on: 14 Nov. 2019.
[41] Online questionnaire developed by the authors. https://docs.google.com/forms/d/19nkoyBZfrmV3WKcKg0qdPKqJWTEEpya0j0RxeU12U-g/edit.
[42] Swarbrooke, J., *Sustainable Tourism Management*, Cabi Publishing: New York, 1999.

CHALLENGES OF COMMUNITY PARTICIPATION IN COMMUNITY-BASED TOURISM IN RURAL AREAS

TSHEPISO J. SETOKOE & TAKALANI RAMUKUMBA
Walter Sisulu University, South Africa

ABSTRACT

Community-based tourism (CBT) has, in recent times, been looked upon as a catalyst for a tourism industry that is sustainable within the rural areas of the developing world. This is seen in its ability to bring about positive change within a rural community inclusive of social justice, empowerment (both economic and social), equity in distribution of benefits and overall community development. These impacts have been seen to justify concerted efforts of tourism development stakeholders in involving local communities, particularly community members in rural areas of the developing world. However, despite the efforts aimed at affording the local communities an opportunity to meaningfully participate in development initiatives aimed at their own development, community participation remains a challenge. The above narrative clearly extrapolates the aim of this particular study as it is intended to establish challenges that the community of Nqileni village in the Eastern Cape province of South Africa are faced with in relation to their participation in the development of tourism and tourism-related initiatives undertaken within the area. The study employed a qualitative research approach coupled with an in-depth exploratory examination of community participation in the development of rural areas through tourism in Nqileni village. This relied on semi-structured interviews using a non-probability purposive sampling technique to collect the empirical data from 25 respondents. The sample of the study comprised of community leadership and general community members in the village. The results of the study indicate that community members felt that their participation in the tourism industry were challenged because of several factors, those being: elitism, factions, greed and corruption, leadership conflict, language and lack of capacity to participate in development initiatives, lack of support from government as well as lack of funding.

Keywords: community participation in tourism, community-based tourism, challenges to community participation, sustainable rural development.

1 INTRODUCTION

The emergence of tourism as an economic activity in the 1960s has seen concerted efforts in the promotion of tourism by governments of countries in both the developed and less developed world [1]. This has seen purpose being placed on tourism development as a catalyst for the creation of opportunities and facilities for income generation and rejuvenation of rural areas [2]. With this kind of attention, the world has seen little recognition given to tourism development within the corridors of power where policymakers and world leaders reside [1]. As a result, scholars have undertaken to embark on research that will ensure the development of a sustainable tourism which has seen an approach geared towards the participation of local communities, thereby creating better opportunities for the local people to accrue greater and more balanced benefits from the development of tourism within their localities [1], [3]. This has seen a continuous appeal for a collaboration of stakeholders including the local community with the aim of obtaining a clear mandate from all interested and affected parties on the desired outcomes of tourism development [4]. Such participation can ensure that the local communities become the focal point of tourism development, as they are part of the tourism product and therefore, their inputs being pivotal to the decision-making process [1], [5].

While considerable effort has been made to ensure community participation in the development of tourism in rural areas, in many instances, this has presented challenges that

WIT Transactions on Ecology and the Environment, Vol 248, © 2020 WIT Press
www.witpress.com, ISSN 1743-3541 (on-line)
doi:10.2495/ST200021

hinder meaningful and capable participation of the citizenry in development initiatives aimed at the emancipation of area and its people [6]. This is seen in the restricted participation or even total omission of local communities in decision-making processes in relation to development initiatives undertaken within their habitat [1], [4]. This study then intends to navigate the maze in which community participation finds itself in relation to the development of tourism in rural areas such as Nqileni village.

2 LITERATURE REVIEW

Community participation is seen as an ideology deeply embedded in beliefs derived from the social and political theories relating to the way society needs to be organised [3]. This gives credence to the importance of establishing connections and sustaining interactions, and in so doing, create strong bonds and relationships, further instilling a sense of belonging, trust and credibility among community members [7]. Similarly, this participation requires that locals be empowered with adequate knowledge about tourism, while enabling them to engage in tourism development [8], [9]. This is further asserted by Hampton [10] who pointed out the importance of communication that will allow for the opinions of the host community to be heard and heeded, while encouraging the development of local small businesses to encourage participation. This can be achieved through placing a requirement on managers and planners to afford residents educational information and initiatives such as workshops and awareness programs [11].

However, the rhetoric surrounding community participation seems to have its roots in an idealistic Western setting, leaving questions around its applicability to settings characterising the developing world. The developing world is faced with constraints in terms of community participation in development initiatives such as tourism, which places barriers on the participation of community in tourism development that go far beyond governance and economics – inclusive of elements such as culture, social norms and behaviour [9], [12].

The sources of constraints to host community participation in tourism development are articulated in the work of Tosun [3], further expanding on Timothy's [12] work, wherein he posits three sources (operational, structural and cultural) of lack of participation in tourism development. Firstly, shortcomings in operational procedures are a result of questionable communication regarding tourism development with destination residents. Secondly, structural constraints relate to institutional factors such as regional power structures, legislative bodies, and economic systems. Thirdly, cultural limitations mostly relate to the capacity of locals in relation to dealing with the impacts associated with the development of tourism. For example, deficient levels of formal education amongst locals may lead to an inability to fully exploit opportunities and benefits that can be brought by tourism development to the destination [9], [13].

The core of sustainable tourism and community development lies in active and meaningful community participation [14]. However, this is often impeded by barriers that disable the intended involvement of the community in development initiatives aimed at improving their quality of lives [13]. Ideally, tourism has the potential to enable meaningful community participation, with the reality being that the barriers exist and impede on the effectiveness of the employment of tourism as tool for community development [14].

The greater challenge faced by literature concerning community participation in tourism sits with the lack of debate by scholars of tourism [13]. This subject is limited to a study conducted by Tosun [3], suggesting that three barriers (operational, structural and cultural) exist in developing countries. According to Aref and Redzuan [14], the barriers formulated by Tosun [3] are meant to facilitate a theoretical understanding of community participation in tourism development without any special reason for the classification.

It must be noted that community participation in developing countries requires a concerted effort involving different ideological beliefs, political forces, administrative arrangements and redistribution of wealth and power [3], [15], [16]. This becomes difficult in a world where community participation is often used as a cloak of words to disguise business as usual – to hide power inequities, gloss over differences, and enable elites to pursue their own agendas [15]. Similarly, this can be a sad result of either the unwillingness of communities to participate due to the absence of ownership sense, insufficient employment generation, deficiency of information, and the unwillingness of top-down institutions to enable real participation [17]–[19].

The challenge faced by the notion of community participation in the development of tourism that is often perpetuated by the reluctance of certain stakeholders with an agenda to maintain a status quo that limits the community's involvement through centralisation of authority, elite domination, poor information sharing and contemptuous professional attitude to capacitating locals with the requisite skills to meaningfully participate [1]. This is a result of poor legal and regulatory systems aimed at supporting and defending the interests of local communities to ensure real participation in the planning, development and management of tourism [3].

Consequently, developing countries like South Africa have to a great extent, been able to achieve a certain degree of success in the legislation of community involvement in development. According to van Niekerk [20], the dawn of a democratic dispensation in South Africa has seen the country becoming exemplary on how, through legislation, the community has been afforded the opportunity to partake in integrated development planning (especially in tourism) following planning practices that were shaped by the history of embedded inequalities that characterised pre 1994 development planning.

The success of the South African community participation project rests upon the decentralisation from national to local government with local government empowered by legislation on the how, when and where the community participation process is administered [21]. van Niekerk [20] makes reference to the uniqueness of South Africa's tourism planning being integrative, goal-oriented and systematic with the overall economic, social and cultural development placed into consideration from the very beginning. This process enables community participation aimed at ensuring success and growth of tourism while providing communities with a real and transparent opportunity to play their part in the development of their own area. This necessitates a need for all stakeholders to work on creating meaningful participation of the community in development initiatives, thereby avoiding the relegation of one group of stakeholders to spectators in development meant to improve their lives [17].

3 SCOPE OF THE STUDY

The scope of the study is confined to determining the challenges faced by rural communities in relation to participation in the development of tourism through the eyes of the residents of Nqileni village, which form part of the four village that that make up the Xhora Mouth Administrative Area on the Wild Coast of the former Transkei, in the Mbhashe Municipality, Eastern Cape, South Africa [22].

According to the Mbhashe Local Municipality's Integrated Development Plan (IDP) [23], Mbashe Local Municipality was constituted in terms of the Municipal Structures Act 17 of 1998 (as amended) and comprises the areas that previously formed the Elliotdale, Willowvale and Dutywa. The Municipality is a Category B, Third Grade municipality situated in the south-eastern parts of the Eastern Cape Province, and is bound to the east by the coastline flowing from the Qhora River in the south to the Mncwasa River in the North along the Indian Ocean. The Municipality is in the North-Eastern part of the Amathole District Municipalities' area of jurisdiction with its main administrative office situated in Dutywa [23] (see Fig. 1 below depicting Mbashe Local Municipality where Nqileni village area is located).

The size of the municipality and its location in relation to South Africa's economic spatial development make the area poor given the socio-economic conditions of the area. This is seen in the high levels of unemployment (estimated at 42%), which is indicated to significantly improved from an estimated 69.9% in 2001 [23]. This is possibly a reflection of the migration out of the area and also probably due to the positive impact of Community Work Programme [24]. According to the municipality's IDP [23], the municipality's economic value is estimated to be R684 million ($43,807,508.64), with agriculture, tourism, community service, construction and trade being the largest economic sectors. Most of the jobs created within the municipality are in the community services sector, which cannot be regarded as sustainable sector for job creation.

Figure 1: Mbashe Local Municipality where Nqileni village area is located.

4 METHODOLOGY

The study employed a qualitative research approach coupled with an in-depth exploratory examination of community participation in the development of rural areas through tourism, in Nqileni village. This relied on semi-structured interviews using a non-probability purposive sampling technique to collect the empirical data required. The sample was chosen based on the respondents' proximity to activities and developments associated with tourism within the village. The study included five semi-structured interviews with community leadership and 19 with the general community members in their different capacities and roles associated with tourism activities in Nqileni village. From a leadership perspective, the respondents were comprised of local headman, ward councillor, ward committee member, manager of Bulungula lodge and local economic development officer for Mbashe Local Municipality whilst community members comprised amongst others, entrepreneurs, employees within tourism sector in the village and general residents of Nqileni village. Data was collected in isiXhosa using a digital recorder. These interviews averaged a duration of 20 minutes per interview. In addition, services of professionals were solicited for the transcription and translation of the data. The data was then analysed using the NVivo version 11, producing themes that allowed for the study to provide a clear picture of the challenges of community participation in community-based tourism in rural areas. This further produced a comprehensive observation report accompanied by nodes, word clouds, cluster analysis, tree maps and words trees, which were used by the research to interpret and present the findings to follow.

5 RESULTS

Emanating from the analysis of the data, five themes emerged indicative of a range of challenges faced by rural communities in terms of participation in community-based tourism in Nqileni village. These are elitism, greed and corruption; leadership conflict; lack of capacity to participate in development initiatives; lack of support from government; and lack of funding and infrastructure. Following is the presentation and discussion of the findings as per responses of the respondents that took part in the study.

5.1 Elitism, factions, greed and corruption

The ugly head of elitism reared its head with respondents indicating the uneven distribution of benefits from development initiatives as these are mostly directed to a few local elites. This is further exhibited in the conduct of community leaders and elites treating development initiatives as personal wealth accumulation exercise, often at the expense of the whole community or the development itself. Such undertakings often compromise the integrity of development initiatives through the creation of factions within the community, which lead to corrupt practice including nepotism, thereby diminishing interest to participate in ordinary members of the community. This is supported by narration from respondents stating the following:

> "No matter how good a development plan might be, there will always be people who will have their own agendas for a particular development. It is therefore up to traditional leaders and general leadership to ensure that they select the right people".
> "They want to make sure that any kind of development there must be something for them to get anything for that particular area".
> "There seems to also be some occurrences of factions between the village chief, local members and business people. This causes segregation and has a negative influence on development".
> "Corruption is also a factor, and this discourages people from participating. Sometimes individuals aim for personal benefit rather than that of the community".

The narratives indicated above give credence to literature critical of the behaviour of elites in rural communities. This is seen in their hesitation to allow ordinary members of the community to obtain a piece of the pie of development initiatives simply because of their social standing, poor or lack of education and lack of connection to the corridors of power [3], [25]. This is attributed to structural limitation of community participation. This is also seen the responses that mirror Mustapha et al. [1] finding indicating the concentration of power in the hands of public administration who see ideas of professionals more valuable than those of local people due to their low levels of education. Consequent to this viewpoint, is the creation of elites, who are then placed in positions of power, thereby leaving community members behind in terms of politics and economic activities [1], [3].

5.2 Leadership conflict

Rural areas in South Africa and many other countries in the world are subject to the challenge of two types of leaderships, namely, traditional and political. Nqileni village is no stranger to the challenge. This has often resulted in strained relations between the two institutions to the detriment of development initiatives, which is often followed by adverse consequence for the very communities that both leaderships claim to care about. These challenges are clearly orated in by respondents as follows:

> "Traditional leaders and ward councillors are not working together in some of the projects. Traditional leader believed that they owned the land, whilst some local communities preferred to work with ward councillors. This then lead to lack of cooperation".
>
> "Because you find out that some of the people or the local communities they prefer to the ward committee… to ward councillor rather for any kind of development that they're interested on. On the other hand, there's a chief here or a traditional leader that says 'We own the land, the land belongs to us. If government or whoever who want to develop this area must come to the chief because the land belongs to the chief.' So, this kind of… kind of not good cooperation between the traditional leaders and the ward councillor".

The above rhetoric asserts truth to the limitations formulated by Tosun [3]. This rings true to operational limitations where a lack of co-ordination between traditional and political leadership is common (Tosun [3]). Such lack of co-ordination hinders community participation in development initiatives and further hampers the destination from developing tourism as a considerable economic activity worthy of investment in time, financial resources and an economic sector capable of creating employment opportunities [1]. It must be noted that the realisation of community participation in development initiatives such as tourism require collective collaboration of all stakeholders including contradictory leadership style of both institutions inclusive of their different ideological outlook in areas such as administration and management of projects together with the distribution of wealth and power for the benefit of the community and the area at large [3], [15], [16], [25].

5.3 Lack of capacity to participate in development initiatives

Respondents indicated a chronic shortage in capacity, disenabling them from meaningfully participating in development initiatives in the area. This is made evident by declarations stating the lack of knowledge, information and understanding of development amongst members of the community. This is further exacerbated by the low levels of education and literacy amongst members of the community coupled, with their inability to communicate in English, as this is the language mostly used in development circles (especially in tourism). This issue is further compounded the lack of interest of the youth in education, and the community's awareness of markets which leaves them ill prepared to exploit opportunities presented by development initiatives such as tourism. These concerns are corroborated by respondents' narrations that follow:

> "It is being uneducated. As I have said that most of us are not educated thus it becomes a barrier to communicate in the English language. We are not able to even converse with the tourists because we do not understand what they are saying".
>
> "Illiteracy remains the biggest challenge. Most of us do not understand the English language and as such, are not able to interact with external people, especially developers and tourists".
>
> "The lack of education among youth is also a challenge. Youth should be educated about what exactly tourism is and how they can play a role in developing it".
>
> "People in the village do not know much about tourism and development and also what they will do with products that can be produced from the area. This makes market intelligence critical".

The above narratives about the community's lack of capacity to participate in development initiatives coincide with the constraints stated in the White Paper for the Development and Promotion of Tourism in South Africa [26], that indicate a need for previously neglected groups (especially in rural areas) to be introduced to community-wide tourism awareness programmes, as well as a wide range of basic skills that will enable participation in development. This is on the back a study conducted by Scheyvens and Momsen [27] indicating that rural areas often start on the back foot, given a multitude of inhibiting factors such as lack of education and skills, limited development opportunities, and limited access to tourism enabling infrastructure and markets.

The findings further give credence to the importance of a community being capacitated for active participation in development activities and initiatives aimed at improving their lives. Efforts to ensure that such capacity is realised, stem from a lack of capacity, which chiefly discourages the enthusiasm of the community to participate in development initiatives, thereby leading to the failure of the community to recognise benefits and ultimately leading to the development being dominated by the elites within the society [4]. This can be remedied through indications made in the work of Murphy [28] as cited in Harun et al. [29] stating the value of such participation as it assists in the decrease of tension often created by development initiatives that are contrary to the aspirations, needs and capabilities of the host community. Consequently, this kind of undertaking can further assist in the improvement of the community's ability to participate in decision-making processes relating to development, and enhance the resources (human, organisational and infrastructural) that will enable the community to draw upon in the quest to improve their own lives [30], [31].

5.4 Lack of support from the government

Respondents indicated a serious lack of government support for the village. This is shown in comments that paint the government as unresponsive to the needs of the village. Respondents further indicate avenues in which government could partner with the community to achieve sustainable development. In addition, they give credit to the establishment of the lodge, and the subsequent development of tourism, for the creation and development of services and infrastructure that the government failed to provide. These are stated in the following respondents' dialogues:

> "Let me just say, we are not getting any of the government benefits. All the developments that that you see here have been brought by establishment of Bulungula lodge in 2004. There is no government involvement whatsoever, in this village".
>
> "I am not sure what those obstacles are because we do everything that is required from us such as voting, but still do not see any efforts from the government to develop our area".
>
> "I mean I think the Eastern Cape government doesn't talk to our communities enough. We tell them that our road is terrible, and they don't do anything about it".

The above rhetoric by the respondents is in sharp contrast with the works of van Niekerk [20] hailing South Africa as a beacon of hope for the community participation arena. This is seen in the country's formulation of legislation (such as Batho Pele meaning *People First*) that places community participation in the centre of development initiatives [21]. This is the legislation that is said to be in place to supposedly afford local communities an opportunity to partake in integrated planning (especially in tourism) following planning processes that were shaped by the history of embedded inequalities that characterised pre-1994 planning [20].

5.5 Lack of funding and infrastructure

Lack of funding and poor infrastructure were also indicated as major obstacles to community participation in development initiatives. Respondents commented on the lack of infrastructure, such as roads, that hinders development within the area. This was further exacerbated by an inadequate telecommunication infrastructure, coupled with power (electricity) shortages and lack of a proper transport system to connect the area with other areas. In addition, respondents further state that the development of tourism has seen opportunities (especially entrepreneurial) become available for residents of Nqileni village, yet there are no financial resources available for them to exploit such opportunities. These problems have placed participation of the community in development initiatives in limbo, and are narrated by respondents as follows:

> "The lack of funds makes it difficult for us to take advantage of development opportunities that arise".
> "Also the lack of financial support affects our progress".
> "Firstly, we do not have roads, electricity, housing and telecommunications".

The above narrations agree with the accounts of the severe lack of infrastructure in rural areas, which limits the participation of rural communities in development initiatives such as tourism [27]. In addition, the absence of adequate transportation services effectively prevents rural communities from participating in the industry, both as potential suppliers of products and services, and as tourists themselves [27]. Consequently, the unavailability of funding and infrastructure disables the core of sustainable development (community participation), which is aimed at improving the lives of the local community [13], [14].

6 CONCLUSION

Community based tourism is key to, and fundamental in, achieving sustainable rural development. Literature has clearly indicated the importance of community participation in development initiatives. This has seen community participation being lauded as the cornerstone of sustainable development [32], especially in tourism, since the preservation and development of resources needed for the development of tourism often represent interests of the local communities and international authorities [33]. This paper highlighted the common problems associated with challenges to community participation in tourism in rural areas such as lack of capacity to participate in development initiatives, lack of support from government, lack of funding and infrastructure as well as elitism, factions, greed and corruption. To ensure maximum community participation, the community should be well informed and educated about tourism initiatives and there should be enough support for them to fully participate in tourism development initiatives in their area.

ACKNOWLEDGEMENTS

The National Research Foundation for financial support and the national government of South Africa for availing funds to assist with the execution of this study.

REFERENCES

[1] Mustapha, N.A., Azman, I. & Ibrahim, Y., Barriers to community participation in tourism development in island destinations. *Journal of Tourism, Hospitality & Culinary Arts*, **5**(1), pp. 102–124, 2013.

[2] Reza Maleki, M., Moradi, E. & Parsa, S., Rural tourism as a way to rural development. *International Journal of Academic Research*, **6**(4), pp. 72–78, 2014.

[3] Tosun, C., Limits to community participation in the tourism development process in developing countries. *Tourism Management*, **21**, pp. 613–633, 2000.
[4] Chili, N.S. & Ngxongo, N.A., Challenges to active community involvement in tourism development at Didima Resort – A case study of Umhlwazini community in Bergville. *African Journal of Hospitality, Tourism & Leisure*, **6**(2), pp. 1–15, 2017.
[5] Choi, H.S.C. & Sirakaya, E., Measuring residents' attitude toward sustainable tourism: development of sustainable tourism attitude scale. *Journal of Travel Research*, **43**(3), pp. 380–394, 2005.
[6] Hendriksen, A., Meaningful stakeholder involvement in decision making processes on sustainability issues. Unpublished Doctoral thesis, Wageningen University, 2019.
[7] Rasoolimanesh, S.M. & Jaafar, M., Sustainable tourism development and residents' perceptions in World Heritage Site destinations. *Asia Pacific Journal of Tourism Research*, **22**(1), pp. 34–48, 2016.
[8] Cole, S., Information and empowerment: the keys to achieving sustainable tourism. *Journal of Sustainable Tourism*, **14**, pp. 629–644, 2006.
[9] Saufi, A., O'Brien, D. & Wilkins, H., Inhibitors to host community participation in sustainable tourism development in developing countries. *Journal of Sustainable Tourism*, **22**(5), pp. 801–820, 2014.
[10] Hampton, M.P., Heritage, local communities and economic development. *Annals of Tourism Research*, **32**(3), pp. 735–759, 2005.
[11] Droga, R. & Gupta, A. Barriers to community participation in tourism development: empirical evidence from a rural destination. *South Asian Journal of Tourism and Heritage*, **5**(1), pp. 129–142, 2012.
[12] Timothy, D.J., Participatory planning: a view of tourism in Indonesia. *Annals of Tourism Research*, **26**(2), pp. 371–391, 2009.
[13] Moscardo, G., *Building Community Capacity for Tourism Development*, Cabi Publication: Australia, 2008.
[14] Aref, F. & Redzuan, M., Barriers to community participation toward tourism development in Shiraz, Iran. *Pakistan Journal of Social Sciences*, **5**(9), pp. 936–940, 2008.
[15] Eversole, R., Remaking participation: Challenges for community development practice. *Community Development Journal*, **47**(1), pp. 29–41, 2010.
[16] Mansuri, G. & Rao, V., *Localizing Development: Does Participation Work?* The World Bank: Washington, DC, 2012.
[17] Gaventa, J., Towards participatory governance: assessing the transformative possibilities. *Participation – From Tyranny to Transformation? Exploring New Approaches to Participation in Development*, eds S. Hickey & G. Mohan, Zed Books: London, pp. 25–41, 2005.
[18] Stone, L.S. & Stone, T.M., Community-based tourism enterprise: Challenges and prospects for community participation; Khama Rhino Sanctuary Trust, Botswana. *Journal of Sustainable Tourism*, **19**(1), pp. 97–114, 2011.
[19] Kala, D. & Bagri, S.C., Barriers to local community participation in tourism development: Evidence from mountainous state Uttarakhand, India. *Tourism: An International Interdisciplinary Journal*, **66**(3), pp. 318–333, 2018.
[20] van Niekerk, M., Advocating community participation and integrated tourism development planning in local destinations: the case of South Africa. *Journal of Destination Marketing and Management*, **3**, pp. 82–84, 2014.

[21] South Africa. 2007. Public Service Commission. Report on the Evaluation of the Implementation of the Batho Pele Principle of Consultation. www.gov.za/sites/www.gov.za/files/eval_implement_consult_0.pdf. Accessed on: 9 Mar. 2017.
[22] Bulungula Incubator. Annual Report. https://bulungulaincubator.wordpress.com/about/. Accessed on: 9 Jan. 2016.
[23] Mbashe Local Municipality. Integrated Development Plan. www.mbhashemun.gov.za/. Accessed on: 13 Jul. 2016.
[24] Bulungula, Bulungula lodge. https://bulungula.com. Accessed on: 9 Jan. 2016.
[25] Setokoe, T.J., Ramukumba, T. & Ferreira, I.W., Community participation in the development of rural areas: a leaders' perspective of tourism. *African Journal of Hospitality, Tourism and Leisure*, **8**(1), pp. 1–15, 2019.
[26] Department of Environmental Affairs and Tourism, South Afica. *White Paper on Development and Promotion of Tourism*, Government Printer: Pretoria, 1996.
[27] Scheyvens, R. & Momsen, J.H., Tourism and poverty reduction: issues for small island states. *Tourism Geographies*, **10**(1), pp. 22–41, 2008.
[28] Murphy, P.E., *Tourism: A Community Approach*, Methuen: New York, 1985.
[29] Harun, H., Hassan, R., Razzaq, A., Rasid, A. & Mustafa, M.Z., Building local capacities towards sustaining community-based tourism development (CBET): experience from Miso Walal Homestay, Kinabantangan Sabah, Malaysia, pp. 1–13. http://eprints.uthm.edu.my/2610/1/Regional_Conf_on_Higher_Edu_Comm_Industry-Engagemnt_2012.pdf. Accessed on: 5 Mar. 2018.
[30] Khosravi, S. & Mohamed, B., Community capacity for conserving natural resources in ecotourist destinations: a review of the concept. *Asia-Pacific Journal of Innovation in Hospitality and Tourism*, **4**(2), pp. 235–256, 2015.
[31] Sithole, N., Ciampiccoli, A. & Jugmohan, S., Community capacity building through tourism: the case of Shakaland Zulu Cultural Village. *African Journal of Hospitality, Tourism and Leisure*, **7**(1), pp. 1–14, 2018.
[32] Lin, D. & Simmons, D., Structured inter-network collaboration: public participation in tourism planning in Southern China. *Tourism Management*, **63**, pp. 315–328, 2017.
[33] Nunkoo, R., Tourism development and trust in local government. *Tourism Management*, **46**, pp. 623–634, 2015.

TAPPING INTO THE UNSUNG POTENTIAL OF TOURIST ATTRACTIONS FOR DEVELOPMENT IN SELECTED MUNICIPALITIES IN UGANDA

CHARLES TUSHABOMWE-KAZOOBA[1], IMELDA KEMEZA[1], MANASSEH TUMUHIMBISE[1], EDWARD ANDAMA[2], RONALD EMOJONG[2], ESTHER NANSAMBA[3] & CHRISTINE AMPUMUZA[3]
[1]Mbarara University of Science and Technology, Uganda
[2]Busitema University, Uganda
[3]Kabale University, Uganda

ABSTRACT

The tourism sector (TS) in Uganda is predominantly nature based with potential to easily affect the fragile ecosystems. The sector is the top exchange earner, employs a significant proportion of the workforce, especially the unskilled, and contributes to the conservation of natural and cultural heritage sites. Therefore, if not managed properly, which must be preceded by documentation, the TS could lead to the destruction of the very resources upon which it depends. The concept of sustainability has often been used to indicate the need to ensure that tourism is developed in such a way that it minimizes the negative impacts. Inspired by the government's current drive to develop tourism cities, in this paper, we analyze the unsung/untouched paradise potential of tourist attractions (UPTA) in three purposively selected municipalities in Uganda. We use the concepts of attractions, activities and sustainability in a broader sense. The results are based on an extensive study of documents, unstructured interviews and observations. Our findings indicate that the three municipalities studied have a variety of UPTA that have potential to contribute to the social, economic, environmental and institutional essentials of sustainability. The UPTA also have a strong multiplier effect of creating jobs for marginalized groups and providing opportunities for designing inclusive institutions and policies. Such findings re-affirm that broadening the scope of sustainability and attractions beyond their physical references re-asserts the importance of UPTA amidst the current criticisms. This means increased livelihoods for not only the youth and women but also adults with high prospects to enhance household incomes. Furthermore, tapping into the UPTA contributes to revenue generation and foreign exchange earnings in the country.

Keywords: attractions, sustainability, unsung tourism, urban tourism, Uganda.

1 INTRODUCTION

Sustainable tourism is the fastest growing sector, dominates economic growth, creates jobs, promotes local culture and products, provides economic growth and shared prosperity world-wide [1]–[3]. Therefore, the expansion of tourism enables rapid economic and social development [4]. Moreover, the sector yields impressive results for the countries, which have been able to exploit their potential tourist attractions (PTA) [5]. The most prominent of these countries in Africa include South Africa, Egypt, Kenya and Tanzania, among others [6]–[8]. Such countries have invested immensely in their hotel industry, game parks, beaches and the transport sectors. Failure by other countries to tap into PTA due to several constraints has resulted in Africa's share of world tourism sector to remain small with the continents tourism growth rate standing at 1.5% in 2000. Despite that, World Tourism Organization's (WTO) long-term forecast for tourism is that international tourist arrivals will top one billion by 2010 and reach 1.56 billion by 2020 if tourist sites and supporting infrastructures in respective African countries are developed [9]. This unprecedented increase in tourist arrivals is an indicator of the importance of tourism to human wellbeing – the human/social dimension of the sustainability. These present great potential for tourism

WIT Transactions on Ecology and the Environment, Vol 248, © 2020 WIT Press
www.witpress.com, ISSN 1743-3541 (on-line)
doi:10.2495/ST200031

to act as a tool for development of both rural and urban places of underdeveloped countries like Uganda to spread tourism benefits to a wide spectrum of stakeholders.

As an example, Uganda has existing PTA that respond directly to growing trends in domestic, regional and international markets. Blessed with abundant rainfall and fabulous climate, the country has extraordinary natural beauty, and is rich in flora and fauna, many of which are unique. Uganda is ranked among the top ten countries in the world in terms of its bio-diversity. Besides, the country has more than 12 forests with large populations of primates in addition to being Africa's number one birding destination with 1056 officially sighted [10]. Various magazines have ranked Uganda as one of the six best places in the world for freshwater fishing [11]. Despite all these, Uganda remains relatively undiscovered as a tourism destination. Specifically, while tourism in the country has developed considerably overtime [12], [13], the municipalities on the other hand have not reaped so much from this profitable and fastest growing economic sector. Instead they have only remained as a stopover for tourists moving to other towns and districts in spite of having UPTA for development. To demonstrate the growth of tourist attractions in the country for example, from 1983 to 2003, the number of tourists increased from 12,786 to 305,720, an increase of 2,291% over 20 years. In 2004, tourist arrivals short to 512,379, an increase of 68% over the previous year [14].

Furthermore, although tourism in major cities of the world is as old as tourism itself, with the increasing number of tourists each year, various countries are focusing on tourism as a tool for developing their cities. Following this trend, Uganda has already named the municipalities of Kabale and Fort portal among others as tourism cities. The role of municipalities in promoting tourism cannot be over emphasized [15]. However, recent studies indicate that if not handled well, cities might face the over tourism crisis [16]. Such crises may even lead to the downfall of the very city which was envisaged to benefit from tourism [17]. Accordingly, this is the gap the authors of this paper are addressing.

Hence, the above area of debate gives rise to the following research guiding objectives:

(i) To document ecologically-sound UPTA from the three selected municipalities in Uganda

(ii) To establish sustainable revenue-generation tourism types through UPTA for
(ii) sustainable tourism management and for community development in areas surrounding the municipalities

(iii) To link UPTA in selected municipalities to the national and regional context for enhanced practical collaboration on conservation management between municipalities and the protected area authorities in adjacent parks and tourism potentials in the country.

(iv) To suggest ways of tapping the UPTA in the three selected municipalities in Uganda.

Furthermore, a number of PTA are part of the seven wonders – Gorilla Trekking, Kazinga Channel, Murchison Falls National Park, Rwenzori Mountains, Kibale Forest National Park, Lake Victoria and Mount Elgon are located in tourism distribution hubs. As such Mbale, Jinja, Kisoro, Ssese Islands and local governments have endeavored to develop the attractions. Accordingly, the districts and communities are able to reap from the flourishing TS and consequently support their development. On the contrary, municipalities have modestly exploited into tapping their UPTA, consequently losing out continuously. It is against this background that this study was undertaken to establish the different UPTA for development. What is most important as well as is to suggest ways of marketing the municipalities as major tourists' circuits. Consequently there will be a multiplier effect in the development in terms of addressing the ever increasing unemployment especially youth at 83%, the static employment in the formal sector outside agriculture averaging 15.1% [18], food insecurity, limited technology uptake, the rapid and extensive rural-urban transition and, disease burden, climate change, gender inequality and inequity [19] among the other challenges of implementing the Sustainable Development Goals (SDGs) in Uganda [20], [21].

The authors of this paper are from three universities. Therefore, in order for universities to continuously provide quality, relevant education and its application to community development, zeroing on the sector is the ideal thing to do. For this reason, the universities agree that tourism and hospitality sector remains the top exchange earner at nearly US$1.6 (about 6 trillion annual). This is almost 23% of Uganda's export earnings. It brings foreign exchange for the development of the economy. Also the sector contributes 9% of the Gross Domestic Product (GDP) and employs 8% of Uganda's workforce. The increasing tourist arrivals into Uganda have been increasing by 21.7% from 2010 and currently are at 1,151,000 visitors [22]. Most important, we point out that this paper prioritizes the private sector particularly the informal business of the sector.

Therefore, given the development potentials of tourism and the likely negative impacts on cities, this paper aims at documenting the UPTA in three Ugandan municipalities: Mbarara, Kabale located in South Western Uganda and Tororo in Eastern Uganda, and recommends measures to sustainably develop these municipalities by minimizing the negative impacts. In the next section we will provide an overview of tourism in the country before providing the conceptual framework, methods and a discussion of the results of our study.

2 LITERATURE REVIEW

Tourism in Uganda began in earnest in the 1950s with the gazetting of Queen Elizabeth National Park, Murchison Falls National Park, Kidepo National Park and other protected area system such as the gazettment of Bwindi Impenetrable Forest (BIF) in October, 1991. The principal attractions were the superabundance of animals and the incredible diversity of scenery. By the 1970s, there were hundreds of thousands of holidaymakers flooding into Uganda and the country was an integral part of the then famous East African Circuit, which also covered Kenya and Tanzania. During the 1960s tourism was the third largest foreign exchange earner for the country, after the traditional cash crops of cotton and coffee, grossing up to US$19 million in 1970 [23]. The sector as with other sectors of the economy, however, suffered tremendously from the years of political turmoil in the period 1970 to 1986. This period saw a major deterioration in the services offered by hotels, neglect of game parks and reserves in addition to the incessant political turmoil, a deterrent to tourism [24].

We point out that Uganda is endowed with plentiful natural and cultural resources. These resources form the basis for a very lucrative tourism industry, create employment and generate income not only to the dwellers in the major economic centers but also in rural areas. Furthermore, through tapping unsung tourism potentials, the country will be better able to protect its rich biodiversity and cultural heritage [25]. A complex amalgam of morphology, climate, fauna and flora make the country one of the popular tourist attractions in the world since the sector is nature based [26], [27]. Additionally, the country boasts of numerous volcanic mountains such as Moroto and Elgon in the East, and Muhavura, Virunga and Visoke in the Southwest. Volcanicity which occurred during the Plio-Pleistocene period is also responsible for over 200 crater lakes in Western Uganda. Many of the crater lakes contain water from Mount Ruwenzori rising to a height of 5,125 m above sea level is the highest point in Uganda and the third highest mountain in East Africa. It has 37 glaciers which are remnants of the Plio-Pleistocene glaciation. Just like above, attractions in Uganda are pointed out including those in western Uganda, but still, the country has unsung attractions which need to be exploited.

The sunny climate of the country is complementary. It has to be emphasized that the elevation moderates the temperatures despite the fact that Uganda lies astride the Equator. The cool temperatures of Kigezi Region have made Uganda to be referred to as "the Switzerland of Africa". Truly, the warm climate is just one of the unexploited attractions in the country. This is one such ignored potential to develop like sunbathing. This therefore indicates a gap that requires an investigation.

Studies indicate that Uganda is second to Australia and has over 107 species of snakes of the 2,700 in the world. Furthermore, there are over 1,010 species of birds [28]. As an example, Queen Elizabeth National Park alone has over 550 species which is an African park record. While these statistics are quite interesting, little is documented about bird watching considered as unsung tourists' attraction as an activity in the selected municipalities. If tapped, however, as a way of urban tourism, it will also be a way of conserving birds. Another great advantage of Uganda as a tourism center is the presence of mountain gorillas in Mgahinga Gorilla National Park and Bwindi Impenetrable National Park. These include many natural equatorial forest reserves such as Kibale, Mabira, Budongo and Maramagambo as well as montane forests on Ruwenzori, Elgon and the Virunga mountains. In many isolated places, there exist giant trees like the Nakayima tree in Mubende. It is 350 years old, 40 m tall with 15 buttresses dividing the base into separate bays or transepts [29].

Sports facilities are also available in the country. Golf players can enjoy themselves at the vast Kampala Golf Course and Lugogo indoor stadium (for indoor games) in Kampala City. Tourists are able to return to their countries of origin with crafts like mats, baskets, drums, bags, pillows and bed covers from cattle hides. These are sold in many outlets such as souvenir and craft shops located in hotels, Entebbe International Airport and along the roads. The peace prevailing in most parts of the country enables the availability of a flourishing and entertaining night life [30]. The selected municipalities have sports facilities which need to be tapped, a gap which this study intended to bridge.

The launching of the Tourism Master Plan (TMP) during the same year and the relative peace and stability, have combined to attract investors into the country. The government also awarded a number of concessions for developing accommodation facilities for tourists in the national parks. As a result, an additional 1,052 new beds of international standards were added by the end of 1995. A significant change that has taken place in the tourism sector is gorilla viewing. It is now the greatest tourist attraction in the country when Bwindi Impenetrable National Park (BINP) was declared a National Park (NP) in 1991 and started gorilla trekking on April 1, 1993. As an example in 1994, 2,476 tourists visited the park [31], [32]. The highlighted general development plans being pursued by the government is not clear how to tap the unsung potential attractions to develop in the municipalities.

Finally, tourism as an industry has many components, an undeveloped and narrow product range, involves inadequate and unskilled manpower/personnel. Along with transportation, it includes such products as accommodations, food and beverages services, shops, entertainment, aesthetics and special events. Therefore, it is rare for one business to provide the variety of activities or facilities tourists need or desire. This adds to the difficulty of maintaining and controlling the quantity and quality of the experience. Thus, to overcome this hurdle, relevant stakeholders involved in the sector related businesses, agencies, and organizations need to work together to package and promote tourism opportunities in their areas and align their efforts to assure consistency in the overall product quantity and quality [33]. Some of these related businesses are untapped.

With the above literature search, we have also established the key words that were being used in connection with tourism. These appeared to fall into five categories. With slight modifications, the groups are natural attractions (NA), recreation and sports facilities attractions (RSF), religious attractions (RA), historical attractions (HA) and hospitality, hospital, University and factories (HHUF) attractions. The same was done with the tourism type. Our examination revealed 18 tourism types: adventure tourism (AT), business tourism (BT), cultural and heritage tourism (ChT), community tourism (CT), cultural trail tourism (CTT), eco-tourism (EcT), faith tourism (FT), mountain climbing tourism (McT), medical tourism (MeT), medicinal tourism (MT), missions and volunteer tourism (MVT), nature-based tourism (NT), regional tourism (ReT), rural tourism (RT), sports tourism (ST), traditional dress code (TDC), township tourism (TT) and wine tourism (WT). We use these abbreviations to represent the tourism categories and the tourism types/activities respectively in the subsequent sections of this paper.

3 CONCEPTUAL FRAMEWORK

In 1972, Clare A. Gunn, as cited by Lew [34], asserted that, "without attractions, there would be no tourism". Also makes the point that without tourism there are no attractions and goes on to summarize the importance of attractions in tourism. Therefore, attraction is our major concept in exploring the UPTA in light of the sustainable tourism debates.

3.1 Attractions

But as Lew [34] states, PTA have been conceptualized in various ways – as exploitable resources, marketable products, and image among others. Current research on PTA advocates for a shift from traditional attractions to Smart Tourism Attractions (STA) by embracing smart technologies that accurately link tourist preferences and destination attributes [35]. Despite the various conceptualizations, our understanding is that there is a seeming consensus that an attraction draws/evokes interest, desire to see, hear, use and feel among others. In our view, this implies that an attraction transcends the tangible (physical) aspects to include the intangible (psychological, spiritual, intellectual and emotional) domains. Once drawn by an attraction, people do a variety of things thus, our second concept of tourism activities/types.

3.2 Activities

Activities refer to the range of things that a tourist does at a place, event, and people/community to which they are drawn. In some cases, activities reflect the interest of tourists and have been used in developing tourist and tourism typologies. For example, ecotourism scholars use activities to differentiate ecotourism from mass tourism. They emphasize that activities have to be small-scale, locally owned, ecologically sensitive and provide learning opportunities [36]. Therefore, activities are at the core of the tourist experience. Campos et al. [37], on the other hand observe that tourism has shifted from mere gazing to active physical, mental and emotional interaction with the nature, other tourists, residents and technologies. Therefore, our analysis considers how the UPTA provide opportunities for such interactions. Most importantly, is that in our analysis we also reflect on how the tourist experience equally creates meaningful experiential moments for the residents, and staff/employees. So far, it is clear that this conceptualization of attractions and activities highlights important linkages with the main tenets of sustainability. Therefore, we explore this concept in detail.

3.3 Sustainability

The generic sustainability model focuses on balancing the economic, environmental and social imperatives taking into consideration the future generations [38]. Important for our study, Chapter 9 of the Brundtland report highlighted the urban challenge for developing countries as increased industrial pollution, population, inadequate poor sanitation facilities, and most importantly underdeveloped land/resources. As a result, one of the recommendations was urban agriculture and promotion of public-private cooperation to utilize these resources. As if to emphasize the recommendation for cooperation, Valentin and Spangenberg [39] adds the institutional imperative that focuses on the mechanisms and processes of inclusive participation in the development process.

Moreover, Deery et al. [40] add that sustainable tourism should also consider the wellbeing of the employees. The recurrence of these issues in the Sustainable Development Goals (SDG) and the relative ability of tourism to contribute to all these imperatives despite the negative impacts of tourism [41] indicates that sustainable tourism debates are still as pertinent today as they were in the 1980s. Based on that importance, our analysis of UPTA explores the tangible and intangible attractions in terms of the possibilities they offer for holistic experiences for the tourists, residents and staff, and how these contribute to the social, economic, environmental and policy/institutional imperatives of sustainability (Fig. 1).

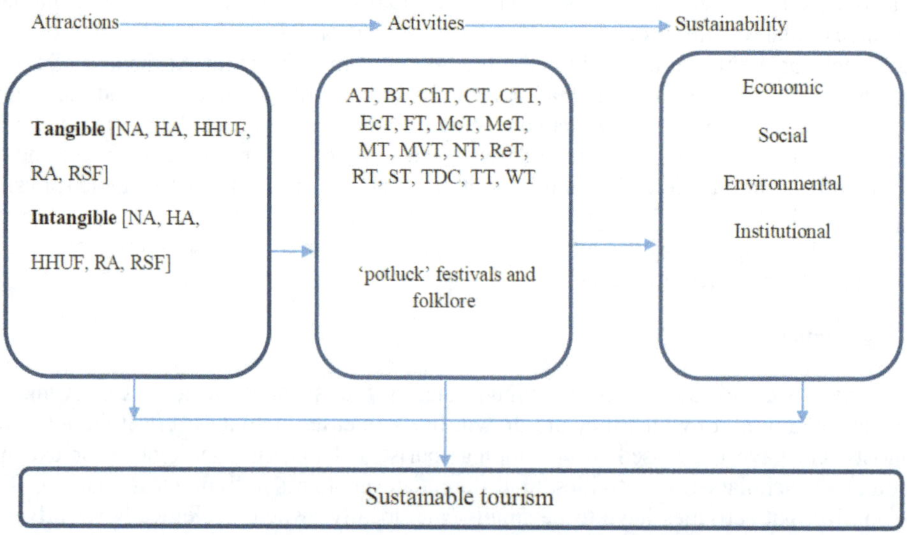

Figure 1: Conceptual framework for analyzing UPTA.

4 METHODOLOGY

The three selected municipalities have unique UPTA. They boast of a fastest growing commercial center with a growing population, leisure, religious and education centers. The study was conducted from September 2019 to March 2020. Ethically, this study posed minimal risk to those who were involved. It was based on generally available information, documents, publications, archival materials and through inventory and in-depth interviews. Each interview was individually tailored to draw out information from which interviewees could freely express their ideas [42] and make comments regarding UPTA of the urban settings and the surrounding areas/communities. Purposive sampling and simple random sampling were used to identify respondents in the data collection process (Table 1).

Table 1: Study participants from the selected municipalities in Uganda.

Name	Gender	Position; number	Qualifications	Years of experience	Data collection tools; sampling procedures
Eastern Uganda					
BU	M/F	TS; 4	TESTS	Various	IS; P
Tourism clusters					
Busoga tourism network	M/F	Members; 5	IB	Various	IS; P
Eastern entrepreneurship and tourism	M/F	Members; 4	IB	Various	IS; P
Tourism associations	M/F	Members; 6	IB	Various	IS; P
Hotel owners and hospitality	M/F	Owner and Manager; 3	IB	Various	IS; P
Crafts associations	M/F	Members; 7	IB	Various	IS; SR
Local governments	M/F	Various; 3	LGSs	Various	IS; P
Tour operators	M/F	Owners; 2	IB	Various	IS; SR
Southwestern Uganda					
MUST	M/F	TS; 3	TESTS	Various	IS; P
KAB	M/F	TS; 4	TESTS	Various	IS; P
Local governments	M/F	Various; 3	LGSs	Various	IS; P
Tourism clusters					
Kigezi	M/F	Members; 5	IB	Various	IS; P
Gantone	M/F	Members; 4	IB	Various	IS; P
Hotel owners and hospitality	M/F	Owner and Manager; 3	IB	Various	IS; P
Tourism associations	M/F	Members; 2	IB	Various	IS; P
Crafts associations	M/F	Members; 5	IB	Various	IS; SR
Training institutions	M/F	TS; 3	TESTS	Various	IS; P
Cultural institutions	M/F	Delegated; 2	IB	Various	IS; P
Religious institutions	M/F	Leaders; 3	IB	Various	IS; P

We used in-depth semi-structured qualitative interviews for data collected which was conducted among 71 study participants (KM – 19(26.7%); MM – 18(25.4%) and TM – 34(47.9%)) as "representative characterization" [43] to gain information on unsung tourism potential in the selected municipalities focusing on the urban tourism development and also to explore the weaknesses and strengths of local authorities in terms of their participation in the tourism planning process in the urban setting.

Additionally, observation method was used and this refers to an extremely wide range of processes carried out, both in a selective, specific, systemic component, and especially at the level of interaction relationships with the target items and recording the information systematically. The semi-structured interviews were recorded and analyzed with the framework techniques. All recordings of the interviews were transcribed verbatim and categorized into keywords, followed by indexing, coding, and classifying by themes and categorization of the data into themes was conducted repeatedly and frequently during the analysis phase.

Both primary and secondary data was collected. This qualitative study is based on a study of various documents such as district and municipal tourism plans, the Kigezi Tourism Cluster Platform baseline survey report on tourism in the region, Uganda Tourism Master plan 2015, Uganda National Development Plan II and III and Uganda's Vision 2040, Ministerial policy statements of the Ministry of Tourism Wildlife and Antiquities (MTWA) and Uganda Wildlife Authority (UWA) policy documents.

Furthermore, 71 unstructured interviews were conducted with the Commissioner Department of Antiquities and Museums under the MTWA, municipality Planners, Commercial Officers, Environmental Officers, selected LC1 chairpersons, Tour and travel operators, owners, Managers/Caretakers of potential attractions, tourists and excursionists found at such attractions located in MM, KM and TM. The documents and unstructured interviews provided insights into the UPTA. This data was augmented by observations aided by field visits to the sites conducted between the months of October 2019 and March 2020.

The potential sites were analyzed using the framework provided in the next section. Data was analyzed and interpreted using both qualitative and quantitative techniques of analysis.

5 PRESENTATION, ANALYSIS AND DISCUSSION OF STUDY FINDINGS

This study sought to document ecologically-sound UPTA and facilities that could be developed into tourist activities (Fig. 2).

We present in Table 2 the key UPTA categories and the tourism types.

Figure 2: Examples of UPTA in selected municipalities in Uganda.

Table 2: Category of UPTA, examples and tourism type in the selected municipalities in Uganda.

Category of UPTA	Key examples	Tourism type
Mbarara Municipality, Southwestern Uganda		
NA	River Rwizi and its associated waterfalls and beautiful sceneries, Round topped Hills, traditional trees, medicinal plants, indigenous flora and fauna species, various bird species	EcT, AT, CT, McT, MT, NbT, RT
RSF	Kakyeka Stadium, Booma Golf Course, numerous hotels	ST, CT
HHUF	Mbarara Regional Referral Hospital (MRRH), informal wine processing, MUST, Nile Breweries, hotels	BT, MeT, MeT, MVT, WT
RA	Churches, Hindu temple, mosques, East Africa Revival Museum,	FT, MVT,
	Amahembe Gente, Burial Place for former colonial administrators, Old House	EcT, ChT,
HA	Burial Site of the former Kings of Ankole, eclipse hill, traditional huts, cultural tours	TDC, ReT, TDC, TT
Kabale Municipality, Southwestern Uganda		
NA	Hills, cultivation terraces, meandering roads, nuclei settlements systems	AT, McT, NbT, RT
RSF	Golf grounds, Kabale Stadium, numerous packaged hotels	ST, CT
HHUF	Kabale Regional Referral Hospital (KRRH), numerous hotels, informal wine processing, KAB, hotels	BT, MeT, MVT, WT
RA	Church and mosque buildings,	FT, MVT
HA	Cultural Institutional infrastructures	ChT, CT, CTT, ReT, TDC, TT
Tororo Municipality, Eastern Uganda		
NA	Tororo Rock (TR), volcanic plug, Indigenous plant species, various bird species, caves, Usukuru hills, cable car, Busitema forest, nuclei settlement systems	AT, EcT, AT, CT McT, MT, NbT, RT
RSF	Golf Course	ST, CT
HHUF	BU, cement factories – Tororo Cement industries ltd, Simba Cement and Hima cement, inland port, Tilda Kibimba rice Scheme	BT, MVT, CT
RA	Church and mosque buildings	FT, MVT
HA	Adhola cultural Institution (ACI), the Bayole, the Etesot and the Bagisu	ChT, CTT, TDC, TT

Source: Field research, September 2019 to March 2020.

5.1 The UPTA in MM

As indicated in Table 3, this study established that "Ebituro bya Abagabe site" is one of the potential attraction that can be developed to attract tourists and recreationists to stay longer in the MM. Located in Nkokonjeru in Kamukuzi Division, this study discovered that although the site which is currently under the Department of Museums and Monuments (DMM) houses prominent former kings of Ankole Kingdom, it is in a sorry state. We indicate that if this site is well-developed, it could attract tourists to the municipality. We suggest the development of cultural trail to link all important cultural attractions in the municipality.

According to the Commissioner of DMM-MTWA, MM has several cultural sites that can be developed by linking them in a trail which provides connectivity of activities to create meaning and experience worth selling to tourists. Such cultural sites include the burial site of the traditional kings of Ankole, the Headquarters of the Abagabe at Kamukuzi, Kiyanja Lake, River Rwizi, indigenous trees and shrubs in the hills of Biharwe division among others all of which could be developed into PTA. This study further indicates that once a cultural trail is developed, cultural and gift shops could also be developed alongside so as to provide an avenue for tourist shops most especially those interested in traditional shoes, necklaces, and drums among others. Besides that, traditional dance could be developed to entertain tourists. We noted that Igongo Cultural Centre is still in its infancy stage in this tourist product. All this could provide alternative activities for tourists and a reason to stay longer than bypassing MM.

The presence of River Rwizi with its magnificent meandering and other features provides an option for boat riding and canoe loving tourists to enjoy scenic viewing, natural plant identification and bird watching among others. Most of the river sections are navigable. This would provide more fun and experience for tourists besides already existing UPTA in the municipality. If the tourist market segment of MVT is developed, the municipality could be a safe landing haven for such.

CT activity was also identified as worth developing. With a shift in tourist interests from mere wildlife safaris to responsible, cultural and community tours, the municipality can position herself very well. To benefit from community tourists should be welcomed in homes of Ugandans so as to participate in the day today routine of the local people for an in-depth experience and understanding of lives of Ugandans. This would not only be exciting but would be a worthwhile activity to occupy the tourists, which means more expenditure.

5.2 The Lake Ngorogoza and its UPTA in KM

In KM, we use the example of Lake Ngorogoza to demonstrate how the municipalities can tap into UPTA (Table 3).

Table 3: Tapping into UPTA for development: The Lake Ngorogoza Case, KM.

Description and location	Tourism type	Contribution to sustainability imperatives
Ngorogoza Lake: This is a man-made pond surrounded by a lush swamp – a habitat for pretty birds such as the hammerkop, sacred ibis, pied kingfisher, malachite kingfisher, black headed, yellow backed and red chested weavers among others. This is in addition to invertebrate and plant diversity undoubtedly important providing a greenery in a predominantly constructed area. The lake named after the famous leader of Kigezi Region who, in spite of lack of formal education was very instrumental in the history of Kigezi as the first general secretary appointed by the colonial administration.	EcT, AT, CT, McT, MT, NbT, RT	*Environmental* – Protection of the biodiversity and all associated benefits to people's health. *Social* – Revitalization of historical and cultural heritage • Fostering community pride in their heritage • Provides learning opportunities for the tourists on nature, culture and history of Kigezi • Enriched tourist, employee and residents experience through learning activities *Economic* – revenue from tourist payments for the activities to the owners, taxes to municipal authorities. *Policy/institutional* – offers an opportunity for various institutional arrangements e.g. Private Community Partnerships or Public – Private Partnerships (PPP). • Can act as a focal point for other community activities by taking into account the activities of other residents around the lake leading to a joint e.g. Mwanjari Tourist Destination Management Team.

5.3 The Tororo Rock and its UPTA in TM

In TM, there is the Tororo Rock (TR). The geographical coordinates of the rock are: 00 41 06N, 34 11 01E (Latitude: 0.6850; Longitude: 34.1836). The rock is generally visible from anywhere in the Tororo District (TD). Thus, it is a major tourist attraction for the district. There is a saying in Tororo that: "The Eiffel Tower is to Paris as the magnificent TR is to TD". The highest altitude of the rock is 4,865 ft (1,483 m) with a gradient of about 0.75 from the east to the west. On the rock's slopes are ancestral caves and various rock paintings. The elevation of Tororo Town is 3,888 ft (1,185 m). Therefore, TR towers 977 ft (298 m) above downtown. Furthermore, the TR is a volcanic plug or a volcanic neck or lava neck, which is a volcanic landform created when magma hardens within a vent on an active volcano. The plug is preserved after the erosion removes the surrounding rock while the erosion-resistant plug remains, producing a distinctive upstanding landform the rock.

This study indicates that the UPTA of TR are many. As an example, the rock has caves some of which act as shrines, home for bats, snakes, and other reptiles. We observed that the rock is an important bat sanctuary in the district. It also has indigenous plants as well as wild mammals such as monkeys, rabbits, birds, porcupines, small antelopes which are all rarely found in other parts of TD due to the degraded natural habitats. Recently *Aloe tororoana* plant which is restricted in occurrence was confirmed to be only found in the two locations (i.e. TR and Usukuru Hills) in TD in the whole of Uganda and perhaps the world. This creates the urgency to conserve the plant as it might disappear from the earth's surface. Andama and Alezuwo [44] indicates that TR is perhaps the only remaining site in the district with more natural or indigenous vegetation. Some of the indigenous plant species still grow intact on the rock while most natural vegetation in other parts of the district has disappeared due to intensive farming and increased human settlement.

5.4 Suggested key strategies for promotion of UPTA

From the empirical evidence of this study, the following are the key strategies for promotion of UPTA in the selected municipalities in Uganda. Some of these strategies are:

1. In TM, promote rock hiking along the trails and rock climbing to the top of the rock. This can be an annual event for the site to be widely advertised.

2. Both MM and TM, tap into viewing the indigenous trees and bird species.

3. In all the municipalities, promote visits by various academic institutions for study
3. purposes, namely, biology, ecology, geography and chemistry.

4. All the municipalities, promote the documented UPTA sites for research activities on ecology and natural resources management in urban setting.

5. In all municipalities, local artisan materials be promoted at the center as commercial business for tourism and local CT.

6. Annual cultural gala during which various cultural groups in the municipality to
6. express their cultural wealth, commonality, such as poetry, song, dance, festivals and forms of art as this promotes UPTA in the country.

6 CONCLUSIONS, RECOMMENDATIONS AND POLICY IMPLICATIONS

This section advances the conclusions, recommendations and policy implications derived from the study findings. The section starts by pointing out the conclusions from the study findings. The section later points to the recommendations and directions for further research, and policy implications that were advanced by the study based on empirical evidence provided by the study findings on tapping UPTA to develop in the selected municipalities.

The findings reveal that despite the fact that the selected municipalities are still unable to compete with most tourist sites and destinations within the country for both international and domestic tourists, they are in position to provide most outdoor recreation and urban tourist attractions once all the potential attractions identified by the study are well developed. We categorized the UPTA in 5 categories especially the intangible ones. Majority of the UPTA observed were found to be in a very sorry state, an indication that there are either no policies or plans to protect, conserve and develop such attractions. There is absence of a Tourism Officer at both Municipalities and District level an indication of disinterest by the district in as far as tourism development is concerned. Moreover, all the three municipalities had no useable developed website. Additionally, the study established that alongside an overlap of tourism types, there is also municipality specific tourism types which once developed could reckon with sustainable revenue generation. Also, UPTA in selected municipalities links to the national and regional context through SDG, Goal 5 on the planned inclusion of marginalized groups and on one of the 8 AU/NEPAD objectives in Tourism Sector, objective 4.

6.1 Recommendations and directions for further research

In this article, we offer several recommendations for tapping into UPTA for development in the selected municipalities in Uganda. Firstly, there is need for the local governments to establish a tourism office/department staffed with qualified tourism officers to document an inventory of UPTA in consultations with other stakeholders involved in the tourism value chain. We suggest that a Municipality Tourism Development Plan (MTDP) should be designed to guide tourism development. Once the MTDP is in place, it will guide all stakeholders on what and how to develop the UPTA. A leaf should be borrowed from Districts in the country which have Tourism Plans that it is a strong driving tool towards the development of the sector. While the Ministry of Tourism, Wildlife and Antiquities (MTWA) is made up of Departments, namely, Tourism Development, Wildlife Conservation, Museum and Monuments, this study indicates there is minimal working with Municipalities to attract private investors to develop the untouchable paradise into viable activities capable of attracting tourists. Besides that, capital to invest in developing these attractions can be mobilized through public-private partnership (P3) with MTWA playing a leading role.

This study calls for proper management and conservation of most UPTA. We point out that a lot of degradation along the banks of river Rwizi in MM, Tororo Rock and Busitema Forest in TM, and Ngorogoza Lake in KM was observed where natural vegetation is being cleared away for the sake of growing crops. The forest and natural vegetation act as home for bird species popular for bird watching activity. We suggest that the District Environmental Officers (DEOs) in conjunction with other stakeholders should ensure the conservation of these natural greens.

On a critical note, there is need for extensive marketing, while focusing and incorporating urban authorities in order to exploit the incredible UPTA in Uganda to be known worldwide. Major players in the country's tourism sector such as the Uganda wildlife Education Center (UWEC), Uganda wildlife Authority (UWA), Export Promotion Board (EPB), Uganda Tourism Board (UTB), Uganda Wildlife Research and Training Institute (UWRTI), Uganda Hotel and Tourism Training Institute (UHTTI) and the Civil Aviation Authority (CAA) among others should work together with urban authorities to promote worldwide awareness about UPTA products in the country. Joint marketing and promotion of UPTA products is highly encouraged.

Moreover, there is need to devote more resources to the marketing and promotion of tourism particularly on the international front where per-capita income is the greatest. Severe loss of market share, tourist foreign exchange and tax revenues has been realized, caused by the continued elimination of tourism promotion by the government. The travel and tourism industry is one of the largest in the world and has experienced stable growth in spite of the likely impacts of the COVID-19 pandemic in the future. And yet, the Uganda government has not invested sufficiently in this sector to make transport industry more reliable, timely and affordable for budget tourists. In most cases the transport sector has been left with the private sector making the industry of transportation very expensive and not affordable to the ordinary tourists across the world.

Also, optimization of the TS is anticipated to address the liquidation in the facilities that had subsequently prevented the tourism industry from realizing its potential. More still, continuous training of the existing and potential guides as well as the staff in the Tourism and Hospitality (TH) sector is very crucial. The well-trained staff ensure quality service delivery to travelers in all departments, hence memorable experiences which lead to repeat tourists. Additionally, partnering with the local media and the non-government organizations (NGOs) is also very important. These help in promoting awareness and a positive image about Uganda as the leading tourist destination offering the unbeaten kind of experience to all travelers.

Building the capacity among the previously neglected informal tourism enterprises and emerging entrepreneurs is also recommended. In addition, the provision of information to visitors in all urban centers that keeps them updated on the security situation help in improving their safety and security while in the country. Furthermore, diversification of tourism products aimed at offering good potential for the development is very important. Diverse tourism products encourage various tourism activities, hence attracting repeat visitors.

Finally, an international visitors exit survey be conducted at departure points which should fall under the tourist high season. This is in order to learn what their experiences were and their exposure to UPTA. What best practices in the UPTA are in place? What is/not working in the country for the development of UPTA? This information is essential for all the relevant players in the TS as they work together to guide its development. It is important to note that UPTA are interconnected with the traditional tourist ones. While the suggested recommendations and direction for further research may not be exhaustive, they point to tapping into UPTA for development in the selected municipalities in the country.

6.2 Limitations of this study

The three (7.7%) selected municipalities that informed this research may not be a good representative of all the municipalities (39) or the many towns (over 156) in the country [45]. Also, the results of this study may not fit directly with a municipality that has metropolitan characteristics such as Kampala Capital City Authority.

6.3 Policy implications

This study links to the 2030 agenda for SDGs Goal 5: Achieving gender equality and empowering all women and girls [46]. Marginalized groups play pivotal roles in the management of unsung tourist activities which is public sector led and private sector driven, implying that the private sector plays a significant role. What is more, marginalized groups possess knowledge about things and have different skills which are handy in the different supply chains of the sector which depends on unskilled manpower. Women's contributions to managing tourism business, however, are seldom acknowledged and their capacity to do so remains unrecognized. This challenge has significant gender dimensions because women and girls, who are increasingly playing an important role in untouched paradise of tourism and hospitality sector, are among the poorest and most disempowered people. One of the reasons for this is the lack of gender-differentiated perspectives, which gap is addressed in this paper.

Therefore, tapping into the fastest growing UPTA for development in selected municipalities in the country addresses the gender inequality and social exclusion. Besides, utilizing the country's natural and cultural heritage endowment by realizing their tourism potential will provide significant benefits to the country. Notwithstanding its early stage of development and contributions, the TS will continue to make dramatic progress as the highest foreign exchange earner. Correspondingly, the sector has a positive spill-over effect on other sectors. The country is now a known international tourism destination of choice, which strongly justifies tapping also into UPTA for the benefits it will provide to marginalized groups in particular and the economy in general. In addition, this paper addresses in particular one of the 8 AU/NEPAD Strategic objective in TS, Objective 4 – Promote research and development [47].

Overall, this study improves the visibility and marketing of UPTA products in the country that are modestly exploited and less commercialized. Moreover, the suggested strategies of strengthening existing and emerging forms of the UPTA will enable the youth, women, service providers and unemployed adults to take part in relevant decision making processes which can influence political power that determines the economy of viable business at a national level. Finally, this paper is in response to the human capital development agendas of National Development Plan II (NDPII) and NDPIII in particular Key Result Area (KRA) 4 – Human capital development in tourism sector, aggressive marketing of tourism products, development of heritage and cultural tourism products in Uganda [48], [49].

ACKNOWLEDGEMENTS

This publication is one of the project outputs of the staff from the three Universities – KAB, BU and MUST when responding to the call, Window 3 – Skills Development Facility (SDF) by Private Sector Foundation Uganda (PSFU). Title of the submitted Extended Concept Note (ECN): Community Based Education, Research and Service – PLUS (COBERS-PLUS) for the Improvement of Unsung Tourism and Hospitality Potential in Uganda. Project Purpose: A sustainable mechanism of tourism development to benefit the communities in the Eastern and Southwestern Uganda is established through Universities-led innovative skills training and public-private partnerships. Overall Goal: Functional tourism based on distinctive local resources is developed in the Eastern and Southwestern Regions of Uganda.

REFERENCES

[1] International Bank for Reconstruction and Development/The World Bank, Tourism: Pacific Possible Background Paper No. 4., 2016. www.worldbank.org/PacificPossible. Accessed on: 27 Mar. 2020.

[2] Christie, I, Fernandes, E., Messrli, H. & Twining-Ward, L., Tourism in Africa: Harnessing Tourism for Growth and Improved Livelihoods. The World Bank, 2019. www.worldbank.org/afr/tourism. Accessed on: 10 Jan. 2020.

[3] World Tourism Organization, UNWTO. UNWTO Global Summit on City Tourism, 2011. http://www.unwto.org. Accessed on: 26 Feb. 2020.

[4] NEPAD. The AU/NEPAD African Action Plan 2010–2015: Advancing Regional and Continental Integration in Africa. New Partnership for Africa (NEPAD), 2009. www.nepad.org. Accessed on 21 Mar. 2020.

[5] Priyanto, S.S., Optimizing the Potential Tourist Attraction Culture in the PuppetMuseum Jakarta, 2018. http://Knepublishing.com/index.php/KnE. Accessed on: 16 Apr. 2020.

[6] Ivanovic, M. & Saayman, M., South Africa calling cultural tourists. *African Journal for Physical, Health Education, Recreation and Dance*, **3**, pp. 138–154, 2013.

[7] The Republic of Kenya, Kenya Tourism Agenda 2018–2022. Ministry of Tourism & Wildlife, Kenya, 2018. www.tourism.go.ke. Accessed on: 18 Apr. 2020.

[8] International Bank for Reconstruction and Development. Economic Assessment of tourism in Kenya: Standing out from the Herd. The World Bank, 2017. www.worldbank.org. Accessed on: 16 Apr. 2020.

[9] UNWTO. International Tourism Highlights, 2019 ed. www.e-unwto.org.doi. Accessed on: 13 Apr. 2020.

[10] Namukobe, J. et al., Traditional plants used for medicinal purposes by local communities around the Northern sector of Kibale National Park, Uganda. *Journal of Ethnopharmacology*, **136**, pp. 236–245, 2011.

[11] Kamuntu, E. & Ndiaye, A., Uganda Tourism sector situation Assessment: Tourism Reawakening Report, Technical Document of the World Bank: Kampala, 2012.

[12] Mwaura, F. & Ssekitoleko, S., Reviewing Uganda's Tourism sector for Economic and Social Upgrading Research Series, No. 91, EPRC, June 2012. www.eprc.or.ug. Accessed on 25 Mar. 2020.

[13] The Republic of Uganda, Bureau of Statistics (UBOS). 2019 Statistical Abstract, 2019, Kampala, Uganda. www.ubos.org. Accessed on: 19 Feb. 2020.

[14] The Republic of Uganda, Uganda Bureau of Statistics (UBOS). 2005 Statistical Abstract, 2005, Kampala, Uganda. www.ubos.org. Accessed on: 19 Feb. 2020.

[15] Cheniki, K. & Baziz, A., Evaluating the ability of local municipalities to promote strong brand image in Algerian cities. *GeoJournal of Tourism and Geosites*, **28**(1), pp. 21–34, 2020. https://doi.org/10.30892/gtg.28102-440.

[16] Godwin, H. The challenge of over tourism. *Responsible Tourism Partnership Working Paper*, **4**, pp. 1–19, 2017.

[17] Seraphin, H. Sheeran, P. & Pilato, M., Over-tourism and the fall of Venice as a destination. *Journal of Destination Marketing and Management*, **9**, pp. 374–376, 2018.

[18] The Republic of Uganda, Uganda Bureau of Statistics (UBOS). 2018 Statistical Abstract, 2018, Kampala, Uganda. www.ubos.org. Accessed on: 19 Feb. 2020.

[19] The Republic of Uganda, Bureau of Statistics (UBOS). 2019 Statistical Abstract, 2019, Kampala, Uganda. www.ubos.org. Accessed on: 19 Feb. 2020.

[20] Uganda Bureau of Statistics (UBOS), The National Labour Force Survey 2016/17 – Main Report, Kampala, Uganda, Jun. 2018. www.ubos.org. Accessed on: 19 Feb. 2020.
[21] Jaiyesimi, R., The challenge of implementing the sustainable development goals in Africa. The way forward. *African Journal of Reproductive Health*, **20**(3), pp. 13–18, 2016.
[22] The Republic of Uganda, Uganda Bureau of Statistics (UBOS). The National Labour Force Survey 2016/17 – Main Report, Kampala, Uganda, Jun. 2018. www.ubos.org. Accessed on: 19 Feb. 2020.
[23] Uganda Tourism profile, Uganda Investment Authority, 2002. www.ugandainvest.go.ug. Accessed on: 5 Feb. 2020.
[24] Uganda Tourism profile, Uganda Investment Authority, 2002. www.ugandainvest.go.ug. Accessed on: 5 Feb. 2020.
[25] The Republic of Uganda, Ministry of Tourism, Trade and Industry (2003). *Tourism Policy for Uganda*, Kampala, Uganda, 2003.
[26] The Republic of Uganda, National Planning Authority (NPA). *The National*
[26] *Development Plan 2011-2015.* Ministry of Finance Planning and Economic Development, Uganda. Kampala, Uganda, 2010.
[27] Namukobe, J. et al., Traditional plants used for medicinal purposes by local communities around the Northern sector of Kibale National Park, Uganda. *Journal of Ethnopharmacology*, **136**, pp. 236–245, 2011.
[28] Briggs, P. & Roberts, A., *Uganda: The Bradt Travel Guide*, IDC House: England, 2013.
[29] Basudde., *Uganda Tourist Industry*, New Vision, 1996.
[30] Kiyingi, K., *Tourism Industry in Exciting Comeback*, New Vision, 1996.
[31] African Wildlife Foundation, *Ecotourism Development Project – Bwindi*
[31] *Impenetrable and Mgahinga Gorilla National Parks, Uganda.* A proposal Submitted to USAID Uganda Mission in Collaboration with Uganda National Parks, 1995.
[32] The Republic of Uganda, *Migration and Tourism Statistics for Busia, Entebbe and*
[32] *Malaba*, Statistical Bulletin No: MT/1/1995, Vol. 1. Statistical Department, Ministry of Finance and Economic Planning, Kampala, Uganda.
[33] Mahoney, Field and MacGregor. Tourism Marketing, Michigan State University, 1997. http://web1.msue.msu.edu/im/modtd/33700082.html. Accessed on: 8 Jan. 2020.
[34] Lew, A.A., A framework of tourist attraction research. *Annals of Tourism Research*, **14**(4), pp. 553–575, 1987.
[35] Wang, X., Li, X.R., Zhen, F. & Zhang, J., How smart is your tourist attraction? Measuring tourist preferences of smart tourism attractions via a FCEM-AHP and IPA approach. *Tourism Management*, **54**, pp. 309–332, 2016.
[36] Holden, A., *Environment and Tourism*, Routledge, 2016.
[37] Campos, A.C., Mendes, J., Valle, P.O.d. & Scott, N., Co-creation of tourist experiences: A literature review. *Current Issues in Tourism*, **21**(4), pp. 369–400, 2018.
[38] Brundtland, G.H., Brundtland report. Our common future. *Comissão Mundial*, **4**(1), pp. 17–25, 1987.
[39] Valentin, A. & Spangenberg, J.H., A guide to community sustainability indicators. *Environmental Impact Assessment Review*, **20**(3), pp. 381–392, 2000.

[40] Deery, M., Jago, L., Harris, C. & Liburd, J., Work life balance for sustainable tourism development. *Collaboration for Sustainable Tourism Development*, pp. 151–167, 2018.
[41] Hall, C.M., Constructing sustainable tourism development: The 2030 agenda and the managerial ecology of sustainable tourism. *Journal of Sustainable Tourism,* **27**(7), pp. 1044–1060, 2019.
[42] Robinson, G., *Methods and Techniques in Human Geography*, Wiley: Chichester, 1998.
[43] Rice, S., Sampling in geography. *Key Methods in Geography*, eds N. Clifford & J. Valentine, Sage: London, pp. 223–244, 2003.
[44] Andama, E. & Alezuwo, C., Tourism Potential in the Northern Uganda, A report submitted to Ministry of Tourism, Wildlife and Antiquities, The Republic of Uganda, 2011.
[45] Uganda Bureau of Statistics, *2019 Statistical Abstract*, Kampala, Uganda, 2019.
[46] UNDP, *Transforming Our World. The 2030 Agenda for Sustainable Development.* A/RES/70/1. www.sustainabledevelopment.un.org. Accessed on: 9 Apr., 2020.
[47] NEPAD, *Africa Union Commission Agenda 2063.* The Africa We Want – A shared Strategic Framework for Inclusive Growth and Sustainable Development: First 10-Year Implementation Plan 2014–2023, 2015. www.un.org/en/africa/osaga/pdf/an/agenda2063-first10yearimplementation.pdf. Accessed on: 8 Apr. 2020.
[48] The Republic of Uganda, National Planning Authority (NPA). NDP II. *The National Development Plan 2015–2020.* Ministry of Finance Planning and Economic Development, Kampala, Uganda.
[49] The Republic of Uganda, National Planning Authority (NPA). NDP III. *The National Development Plan 2021–2025.* Ministry of Finance Planning and Economic Development, Kampala, Uganda, 2020.

ALTERNATIVE VIEW OF CROATIAN TOURISM

MAURO DUJMOVIĆ & ALJOŠA VITASOVIĆ
Faculty of Economics and Tourism, Juraj Dobrila University of Pula, Croatia

ABSTRACT

An excessive increase of tourism capacities at the expense of living space and the normal way of life of the local population has been the dominant feature of Croatian tourism development for the last decade. Croatian tourism is all about income generated from tourism development and it is impossible to find any information related to expenses generated by tourism. Without a clever approach to these issues the real price could be paid in the years to come. The construction of newer and newer rooms, apartments and holiday houses disfigures our settlements and environments and contributes to the inflow of tourists, which in turn leads to less interest, less attractivity and, consequently, a fall in prices of goods and services. Under such circumstances, tourists and the tourism industry exploit the local culture, resources, attractions and the whole way of life of the local community. Croatia is in desperate need of a smaller scale of tourism which will not consume but instead preserve the environment and resources and improve the quality of life of the local population and which will not endanger local traditions and the demographic structure of the territory. The main point is that tourism alone is not enough and that the country's competitiveness should not be grounded only in tourism and traditional industries. The country's sustainable development should be based on new technologies, software production, creative industries with high added value, energetics and knowledge and a creative economy in general and Croatia should adapt to new trends and the new age as soon as possible.

Keywords: tourism, tourists, commercialised hospitality, neo-colonialism, consumerism, leisure, authentic, tourism development, local community.

1 INTRODUCTION

Tourism is the largest contemporary economic phenomenon. According to the UNWTO [1] tourism is the third biggest export industry after oil and chemical industries and it accounts for 7% of the overall world export and 30% of all export services. Tourism is a key element of national development strategies and a very important tool in hands of many governments as the initiator of economic renovation. However, tourism is much more than this. It is a complex activity including much more than the commonly simplified understanding of this industry as a relatively simple source of revenue.

One of the significant characteristics of Croatian tourism is the marked concentration of tourist facilities along the Adriatic coast, which indicates that the tourism industry is more developed along the coast than in the continental parts of the country. Another prominent feature of Croatian tourism is its seasonal nature [2]. The first idea crossing everyone's mind when they hear the word tourism is profit, private accommodation providers, big hotels and tourism companies they all count money generated from tourism related services and products and overnight stays. In Croatia everybody talks only about the income from tourism and nobody has ever measured the costs generated by a tourist season. Additionally, excessive increase of tourism capacities at the expense of living space and the normal way of life of the local population has been the dominant feature of Croatian tourism development for the whole decade and without a wise approach to the solution of these issues the real price could be paid in the years to come. According to all available indicators so far Croatia has definitely entered into a period of tourism stagnation and due to a whole array of factors (already widely discussed in Croatian media like the return of Turkey and North African rivals to the tourism market, economic recession in the countries which are the generators of the tourism arrivals etc.) the times of hyper-demand have gone by. The last tourist season has

WIT Transactions on Ecology and the Environment, Vol 248, © 2020 WIT Press
www.witpress.com, ISSN 1743-3541 (on-line)
doi:10.2495/ST200041

also accentuated the continuation of the trend in the growth of private accommodation at the Croatian coast. What is problematic within this context is that there is no reversal of this trend, although it has become clear that the offer has outgrown the demand and it will certainly be more and more difficult to rent out all existing and newly built accommodation facilities in the years to come.

The fact that Croatia has already succumbed to the consequences of the excessive growth of tourism is best illustrated by the following figures related to occupancy rates during the last tourist season when all accommodation capacities were less occupied than a year or two before. During the first eleven months of 2019 hotels' average occupancy rate was 42.6% (0.6% less than the year before) and the same is valid for camps and agritourism and farm households and private accommodation in Croatia. In the last two years the number of beds in private accommodation has increased by 14.8% and the number of overnight stays by 9.3%, which indicates that the rise of capacities is not matched by the rise of demand. In the same period commercial accommodation capacities in Croatia have increased by 10.5% and the number of overnights amounts to a half, that is only 5.7%. The biggest number of overnight stays has been recorded in Dubrovnik, Rovinj, Poreč and Split, where local inhabitants have expressed their dissatisfaction with the poor quality of life during the peak season. In the cities packed with apartments, like Split, only 5% of local inhabitants rent out their accommodation, so the rest of 95% of citizens righteously ask themselves: "what do we have of it?" Late bookings and cancelled reservations were a dominant feature of the last spring and made all those depending on tourism really nervous in addition to wet May and a part of June with 50 rainy days, which just pointed to the extent to which the Croatian tourism is dependent on weather and climate [3].

We bear witness to some other, mainly negative and very troubling occurrences like the street fights between local merchants and innkeepers all over the Croatian coast, the increase of physical violence, racial and gender discrimination of tourists, unreal increase of prices of goods and services and the deterioration of their quality, the increase in parking fees, the expensive umbrella and deckchair rental at the beaches and numerous cases of insults and threats directed towards visitors, the lack of skilled labour, considerable incompatibility between the local infrastructure and the intensity of the tourist arrivals (waste disposal problems, sewage and faecal pollution, etc.) on account of which Croatian tourism has gained a bad reputation. This is not good in long-term. The provision of accommodation has become the local community's way of life, neglecting the fact that tourism is highly unstable product susceptible to seasonality and unpredictable exterior impacts, suffering from heterogeneity of tourism motivations and expectations and very elastic in terms of price and revenue.

What is true beyond any doubt is that tourism includes people, tourists who enter into interactions with other tourists, destinations, cultures and lifestyles. Today one out of every seven people in the world is a tourist, which makes tourism primarily a social and a cultural phenomenon giving rise to many changes in the destinations where it evolves. Tourism is a socially structured activity and the mere nature of tourism expenditure depends upon the socio-cultural determinants (the amount of personal income, age, educational level, personal lifestyle, class membership etc.), which significantly influence motivations, frequency, duration, selection of a destination and a type of travel. It is evident that tourism does not only include the number of achieved arrivals and overnight stays in a destination, which is particularly pronounced in Croatia because of the excessive imperative to accumulate money from tourism, which accounts to one-sixth of the Croatian gross domestic product [3] and because it is the only remaining relevant industry in the Republic of Croatia. Tourism should be studied and researched as an integral social and economic phenomenon owing its popularity to modern way of life and intensive technological advancement resulting in positive as well as negative consequences for the destinations in which it takes place.

This article is the authors' independent attempt to review the contemporary nature of Croatian tourism and trends from the sociological point of view. This critical analysis is based on information collected from various resources and sites including publications of the Institute of Economics in Zagreb, Croatian National Tourist Board's publications and well established sociological literature, which have all been used by the authors to identify the most important social and cultural features or consequences related to the development of contemporary tourism in the Republic of Croatia. The next section describes the relation between tourism and consumerism in order to show that tourism has betrayed its social and cultural goals. Instead of being a vehicle for the promotion of international peace and greater intercultural understanding it has been turned into commercialised activity diluting local culture and transforming it from a process into a product offered and sold in the tourism market.

2 TOURISM AND CONSUMERISM

Since the beginning, tourism has been exploiting people's need, that is, wish for travelling. This exploitation of human curiosity and urge to see and experience something new and different in comparison to everyday life represents the crucial moment when it comes to the development of tourism as a commercial activity [4]. Today tourism has been transformed into the holiday industry and it has been focused on the production and fulfilment of people's wishes and as such it has become a very important and inevitable part of contemporary consumer behaviour [5]. At a symbolic level tourism has become a very important field for the creation of individual identities and by participating in various types of tourism and by travelling to different parts of the world people create their own identities and in this way they send certain messages about themselves to the society [6]. Within this framework tourism functions as a positional good and a proof of someone's prestige and status within the society and it has become a subject of social differentiation and distinction among people [7]–[9].

On the other hand consumer capitalism has brought to the situation in which it has become possible to set price for almost everything that can be produced and consumed (the so called mercantilization of reality). When it comes to tourism it is evident that spaces, history and attractions as aspects of culture have been transformed into goods and placed on the tourism market. Tourism commodifies cultural values and assets of receptive communities and places them on the market as products to be sold and consumed. In this process of packaging and creating a recognisable tourism products and services local culture is trivialised and modified in order to meet tourists' tastes [10]. Culture commodified in such a way becomes devoid of its original character, meaning and authenticity and transformed into commodified attractions, spaces, gatherings and souvenirs provided only for tourists' consumption [11]. Commodification turns tourism destinations into the so called consumeristically structured spaces and tourists accept the role assigned to them by the tourism industry [6]. Such a commodification of space often results in the increased living costs and ghettoised forms of life, which in turn leads to the transformation of destinations into thematic parks and open-air museums (Venice, Barcelona, Dubrovnik…) devoid of authentic forms of social life and abandoned by residents. Tourism is the extension of the consumer society and it is impossible to escape it. In Venice it has become impossible to buy an authentic souvenir because it has become cheaper to manufacture it and import it from China. In Croatia for example once upon a time children used to sell shells from the Adriatic sea as souvenirs and today these shells have been imported from some other seas and oceans.

The notion that it is necessary to alter the awareness about the importance of the authenticity of local souvenirs has been present in the public discourse for the last twenty years. The same is also true for the local gastronomy but it is still very hard to come across authentic, original souvenirs or restaurants offering authentic meals. There is a mushrooming of fast-food facilities, pizzerias, fast and easy digestible food, seasonal restaurants with Patagonian squid on the menu all over Croatian coast. If you go for a walk along a waterfront, you will not experience the real life of a destination because the majority of tourist attractions and the offer of excursions, daily trips and tours is usually packed in the so called front regions [11]. Those who would like to experience and have a taste of a destination's real life should get off the beaten path and move away from main tourist streets into the so called back regions in order to see how domestic population really lives and runs after their daily duties and obligations. Unfortunately, authentic life, local culture and traditions have been removed from the front regions dominated by economic imperatives.

Next section elaborates the complex notion of the changing nature of tourism by briefly analysing some prominent features of the contemporary Croatian tourism offer bringing to the forefront the issue of authenticity.

3 TOURISM AS COMMERCIALIZED HOSPITALITY

In destinations dependable upon tourism the imperative to earn a living from tourism is the strongest motivation. Different cultural forms (carnivals, arts, crafts, religious events and historical re-enactments) have been altered as a consequence of commodification and very often trivialised and mass produced for the needs of the tourism market. In such a process, destinations tend to neglect all traditional aspects of their own culture and ways of life in order to obtain as much money as possible from tourists and that is why cultural events have been transformed into staged spectacles devoid of any cultural meaning [12]. It is a part of the so called commercialized hospitality which has been turned into servility because the guest–host relationship is dominated by the urge of the hosts to attain as much profit as possible and in this way the potential friendship has been turned into a business venture [13].

The race for profit is ruthless and it is the main motivation behind contemporary tourism development. However, everything does not have to look that bleak and commodification does not necessarily have to dilute local culture. There are many theorists [14], [15] who think that the commodification of material culture for the purposes of tourism consumption, that is, the trivialisation of cultural events, festivals, artisan fairs, gastronomy, thematic parks and staging of local traditional culture, etc. does not necessarily have to be a negative phenomenon and that it may contribute to the raise of awareness of the local population about the value and importance of their own culture and its preservation. Obviously, this phenomenon may have positive effects, however, the issue of authenticity has to be taken into account at this point as well. When it comes to Croatian coast it could be said that besides the dominant existential impacts, tourism has definitely contributed to the raise of awareness of the local population about the importance of their own culture and heritage in the sense that the inherited property should not be sold but renovated and this is a positive trend. However, the way in which this property is renovated is problematic. Modern architecture dominates and there is a proliferation of swimming pools in order to meet tourists' tastes and preferences and the villages have been turned into thematic parks lacking permanent residents during winter months and during the off season. The host and his way of life have been excluded from that kind of offer and in long term this leads to depopulation of rural areas and the extinction of authentic life on account of tourism development. In Croatia this kind of offer has already outgrown the demand and to be honest a swimming pool does not prolong the tourist season beyond summer months. A vineyard or a garden could represent an additional value in Croatian tourism as well. Instead of a swimming pool guests could be offered home grown food, preserved nature, the scent of freshly mowed grass, the sound of silence, authentic and organic food, they could taste local cuisine and become familiar with domestic animals. Emotions should be delivered to tourists and they should indulge in local traditions and local specialities and that is why it has become indispensable to work on the raise of awareness and change of local people's mentality. It could be the right path to a more efficient tourism valorisation of the continental part of Croatia and in this way the development of tourism could contribute to diminishing the differences between underdeveloped interior of the country and developed coastline.

Croatian tourism today has been suffering from low levels of loyalty with fewer tourists coming back to a destination already visited sometime before. People do not come because of accommodation facilities but because of natural beauties and the quality of tourism offer. Croatia is and will certainly continue to be sun and sea holiday destination and it should not be ashamed of it. However, beyond economic, tourism must be more versatile and richer in content in order to provide visitors with productive, useful, cultural and creative holidays and enable them to fulfil their creative skills and interests and enhance their personality [16]. Therefore, it is necessary to develop special interest tourism in order to improve and enrich mass tourism, to extend its structure and offer. Despite of all its obstacles and difficulties tourism has a chance to contribute to a balanced regional development and has a huge potential to bridge regional economic inequalities. It is certainly very difficult and time consuming to achieve this goal, however, without more active inclusion of state and local authorities its achievement may even become unattainable.

Before the concluding remarks the nature of large-scale tourism development is debated in the next two sections in order to reveal the most common changes in a destination's economic and social structure as a consequence of over-dependence on tourism.

4 TOURISM AS NEO-COLONIALISM

In spite of all the above mentioned, from the developmental point of view it could be concluded that the massive character, mobility and geographical diffusion are the most valuable features of contemporary tourism, whilst its seasonality, that is, its concentration in space and time are its main negative features and the main cause of massivity and uniformity of tourist traffic [16]. Mass tourism has become the dominant form of development after 1950s everywhere in the world. In the context of such a development and owing to the already mentioned mercantilization of reality (everything is on sale in the tourism market) everything is planned in advance, there is no spontaneity and everything is subjected to the economic viability of tourism [15].

Economically developed countries are the main generators of tourist arrivals accounting for the biggest number of tourists. In such circumstances the development of tourism is under the control and influence of foreign investors and foreign capital draining out almost all income generated from tourism (money leakage). In the accommodation sector, for example, French Accor group manages and owns more than half a million hotel rooms worldwide whilst Marriot International over 1.1 million rooms [17]. Such a development does not meet the interests of the local community and it is totally adapted to suit tourists' needs. This is particularly marked in developing countries but even Mediterranean destinations similar to Croatia, such as Spain and Greece have, in a sense, become colonised by their northern European neighbours during the summer months with their economies greatly dependent on tourism income [18]. In any case, there is always some sort of economic dependency present at all levels of the development of tourism always resulting in a certain degree of inequality between local people and tourists.

Within the framework of tourist-host relationship the incoming culture is superior in every possible aspect in which process the economic aspect is mostly pronounced. The quantity of this superiority is dependable upon the so called cultural distance, that is, the similarities or differences between the tourist and the host [19]. These differences are most obvious in smaller destinations giving rise to huge differences in the standard of living and lifestyles between the tourist and the host and are less obvious in big, metropolitan centres where the similarities between the tourist and the host are bigger. The feeling of inferiority present in the members of the local community is maybe most obvious when it comes to people employed in low wage jobs in tourism industry. Their job is to be always at tourists' disposal resulting in feelings of antagonism, intolerance and envy towards tourists as a consequence of the so called demonstration effect, a process in which the members of the local community try to imitate the lifestyle and purchasing power of incoming tourists [20]. Local people become aware of tourists' affluence usually beyond their reach and they want to achieve a similar level of affluence for themselves. Although, at a basic level, this demonstration effect results in the introduction of alien values and Western ways of life into relatively traditional societies and may lead to increasing polarisation between the older and younger generations, its wider social implications should not be underestimated. Tourism changes the economic and social structure not only of underdeveloped or developing countries but of developed countries as well. The demonstration effect contributes to the abandonment of agricultural production and local arts and traditional crafts because people think that tourism is the easiest way to make money and earn a living. In such circumstances the tourist is not interested in the interaction with local people and local culture and this sort of tourism development inevitably leads to the violation of social balance and does not promote international peace and understanding as a basis for reaching a greater level of respect and confidence among all the peoples of the world [21].

In the last fifty years the development of mass tourism has brought to the changes in the social structure of host societies and instead of the reasonable small-scale tourism development, we bear witness to tourism development beyond any control, which represents a threat to a destination's economic, physical and social carrying capacity [22]. As soon as the threshold of local people's tolerance is exceeded, which has become the case in Croatia in the last few years, the attitude of the local community towards tourism development becomes antagonistic. According to a recent survey the town of Split, at the Croatian south coast, has become infamous regarding the degree of intolerance towards tourists and tourism. 20% of its citizens have expressed their wish to be able to live in a real town as it once was, not in a tourist destination into which it has been turned today [23]. When local people believe and feel that their everyday life has become difficult to conduct on account of overtourism, tourism becomes a problem. Local people should be involved in the tourism development and the creation of the tourism offer as well as in destination branding. The members of the local community must benefit from tourism development and it should be directed towards the improvement of the quality of life of the local population. No wonder that a negative attitude towards tourists has been evoked in local population, especially in those people who have no direct economic benefit from such a tourism development. Popular destinations like Barcelona or Venice are no exception to his rule. Venice has announced the introduction of the sale of entrance tickets, citizens have established their own associations, they have been throwing stones at buses, demonstrating during the arrival of cruisers into the port, etc. All these problems are linked with the endangerment of the carrying capacity of space, that is, the maximum number of person who may dwell in an area without ruining the quality of life in that particular area. Naturally, tourists feel uncomfortable in such circumstances as well. Therefore, it is necessary to establish a balance between the tourism infrastructure and a growing number of tourist arrivals. It is the only path to a balanced tourism development. Reasonable and balanced contact between tourists and hosts is possible only in cases when the local community retains the control over tourism development, in which process all the members of local community, not only those who directly benefit from tourism, should be involved in tourism planning, management and development [19].

Unfortunately, the problems have been swept under the mat or randomly dealt with not only in Croatia but elsewhere as well. Under such circumstances, balanced and harmonious tourist-host relationships are likely to occur only when the tourism product is small scale, locally owned and controlled, and not the major source of revenue and employment for the local community. When hosts become dependent on tourists and outside organisations their way of life and local culture are turned into a commodity. The profits of tour operators and other organisations acquired in receptive destinations represent a form of exploitation and tourists and tourism industry are turned into the exploiters of local culture, resources, attractions and the whole way of life of the local community [22]. Croatia is in desperate need of small-scale tourism, which does not consume but properly valorises its space and resources and brings better quality of life to the local people without endangering traditions, customs and a demographic structure of the territory.

5 TOURISM AS THE ONLY FORM OF ECONOMIC DEVELOPMENT

Tourism is a low profit industry because it takes long to obtain a return from an investment into tourism industry and it brings less added value, especially when it comes to tourism development in underdeveloped and developing countries. In these destinations, and primarily because of their underdevelopment, tourism represents an ideal type of economic development and in most cases tourism is a relatively new activity which in short time has acquired big proportions resulting in high pressure on local infrastructure and resources. In such circumstances local producers and suppliers, agricultural and food industry cannot cope with the increase in demand for their products and services, which have to be imported. Another issue is that it is more lucrative to import than to manufacture because imported goods are cheaper on account of the emergence of the global market and the processes of liberalisation and deregulation, especially in countries like Croatia, where the government fails to protect local production and distribution. Besides, interest and influential lobby and political groups make considerable amounts of money out of food import. Such anomalies are likely to happen anywhere if the intensity of tourism growth reaches high levels. It is unbelievable that we have failed to realise the fatality of nation's economic dependence on tourism. Even popular Mediterranean tourist countries like Spain, Greece and Italy are faced with different problems on account of this dependency, namely, erosion of the social fabric, higher crime rates, waste disposal problems, water supply problems, electric power supply problems, chaos and traffic jams, uncontrolled exploitation of natural and other resources, import of foreign labour as well as economic crises of bigger or lesser intensity. According to the Croatian Tourist Board 92,000 people are currently employed in tourism industry in Croatia [3]. Although this figure is rather low, Croatia is one of rare tourist countries with 116,000 unemployed people suffering from a dramatic lack of domestic labour.

The main point is that tourism is not enough and Croatia must not build its competitiveness solely on tourism. According to the UNWTO [1] tourism industry accounted for 19.6% of the Croatian gross domestic product in 2018, making Croatia vulnerable to every anomaly in the market, which can result in fatal consequences for the state budget. At the same time this is a proof that Croatia has ruined its own industry and food production, so that tourists visiting Croatia consume imported goods and agricultural products. Tourism, for example, accounts for 1.1% of the German gross domestic product, 1.6% of the Great Britain's gross domestic product, 1.9% of the French gross domestic product, 2.2% of the Italian gross domestic product, 4.7% of the Spanish gross domestic product, 4.9% of the Austrian gross domestic product and 5.9% of the Slovenian gross domestic product. Besides Croatia, Malta with 14% and Cyprus with 12.8% have the biggest share of tourism in the national gross domestic product. It means that other sectors of economy should be developed as well. Croatia shall become an economically developed country when the share of tourism in GDP falls below 5%. In economically developed countries tourism is a secondary activity, the cherry on top of the cake. Obviously, Croatia has not learned anything from the process of postindustrialisation. Traditional industry was ruined and new industries and economic activities have not been developed so far.

Tourism alone is not enough and that the country's competitiveness may not be grounded only in tourism and traditional industries. Country's sustainable development should be based on new technologies, software production, creative industries with high added value, energetics and knowledge and creative economy in general and Croatia should adapt to new trends and new age as soon as possible.

6 CONCLUSION

On account of more intensive technological development, improvement of the means of transport and the establishment of the global travelling infrastructure, the development of new media and information and communication technologies, the rise in the standard of living and the increase of free time, tourism has been rapidly expanding today. Technological advancement and access to information has accelerated the globalisation of the tourism industry and enabled greater flexibility for individual consumers, which has contributed to the creation of new needs and the growth of personalised and customised products and services and brought to the development of new types of tourism, which in most cases may not substitute, but may enrich the already existing forms of tourism [16].

Paradoxically, at the same time this type of tourism development has also lead to a further expansion of mass tourism and it has strengthen the effects of deregulation of the air transport, package arrangements and massive production of services. In other words, regardless of the greater liberalisation of the air transport, the increase and development of new travel services (on-line reservations, on-line financial transactions etc.) the majority of tourists are still conventional or mass tourists visiting popular destinations and staying in conventional accommodation facilities and participating in usual activities during leisure time. Majority of them are still buying their holidays from tourist agencies, tour operators and through package arrangements. Therefore, tourism, to the greater part, is still conventional, massive and it will remain so in future as well, although the trends in conducting leisure and the choice of destinations constantly change.

Tourism should be liberated from commercial chains and it should be protected against the imperatives of the consumer society and tailored to fit men and human measure as soon as possible. Tourism should include those entrepreneurs, citizens, families who have a long-term interest and who plan to be inherited by the children of their children. Croatia has many capable and educated people, however, we bear witness to the brain-drain and people leave the country in search of better life and also because of pessimism, depression, hopelessness, and because our society does not offer a vision of better future. Croatia is in urgent need of the change in mentality and greater responsibility of the state and its institutions. The government must establish a legislative framework, create a strategy and encourage reforms and changes. The main point is that tourism alone is not enough and that the country's competitiveness may not be grounded only in tourism and traditional industries. The country's sustainable development should be based on new technologies, software production, creative industries with high added value, energetics and knowledge and creative economy in general and Croatia should adapt to new trends and new age as soon as possible. If not Croatia will face an almost dystopian vision of future in which tourism will continue to be the main branch of economy and Croatia will become a country of underpaid seasonal labour force performing precarious work.

REFERENCES

[1] World Tourism Organization, International Tourism Highlights, 2019 ed., UNWTO, Madrid. https://doi.org/10.18111/9789284421152.

[2] Vukonić, B., The "new old" tourist destination: Croatia. *Mediterranean Tourism: Facets of Socioeconomic Development and Cultural Change*. eds Y. Apostolopoulos, L. Leontidou & P. Loukissas, Routledge: London and New York, pp. 64–72, 2001.

[3] Rašić, I., *Sektorske analize: Turizam. Ekonomski institut*, Zagreb: Zagreb, 2019.

[4] Cohen, E., Who is a Tourist? A conceptual clarification. https://doi.org/10.1111/j.1467-954X.1974.tb00507.

[5] Featherstone, M., *Consumer Culture and Postmodernism*, Sage: London, 1991.

[6] Hannam, K. & Knox. D., *Understanding Tourism*, Sage: London, 2010.

[7] Campbell, C., *The Romantic Ethic and the Spirit of Modern Consumerism*, Blackwell: Oxford, 1987.

[8] Bourdieu, P., *Distinction: A Social Critique of Judgement of Taste*, Routledge and Kegan Paul: London, 1984.

[9] Holden, A., *Tourism Studies and the Social Sciences*, Routledge: Abingdon, 2005.

[10] Smith, M., Macleod, N. & Hart Robertson, M., *Key Concepts in Tourist Studies*, Sage: London, 2010.

[11] MacCannell, D., *The Tourist: A New Theory of the Leisure Class*, London: University of California Press, 1976/1999.

[12] Greenwood, D., Culture by the pound: An anthropological perspective on tourism as cultural comoditisation. *Hosts and Guests: The Anthropology of Tourism*, ed. V. Smith, Blackwell: Oxford, pp. 171–185, 1989.

[13] Turner, L. & Ash, J., *The Golden Hordes*, Constable: London, 1975.

[14] Cohen, E., Authenticity and commodification in tourism. *Annals of Tourism Research*, **15**(3), pp. 371–386, 1988.

[15] Lash, S. & Urry, J., *Economies of Signs and Space*, Sage: London, 1994.

[16] Jadrešić, V., *Turizam u interdisciplinarnoj teoriji i primjeni*, Školska knjiga: Zagreb, 2001.

[17] Statista. International hotel groups by number of hotel rooms worldwide as of April 2016. www.statista.com/statistics/245690/number-of-hotel-rooms-of-international-hotel-groups/. Accessed on: 20 Feb. 2020.

[18] Sharpley, R., *Tourism, Tourists and Society*, Routledge: London and New York, 2018.

[19] Wearing, S., Stevenson, D. & Young, T., *Tourist Cultures: Identity, Place and the Traveller*, Sage: London, 2010.

[20] Urry, J., *The Tourist Gaze*, Sage Publications: London, 1990/2002.

[21] WTO Manila Declaration on World Tourism, 1980. www.univeur.org/cuebc/downloads/PDF%20carte/65.%20Manila.PDF. Accessed on: 3 Mar. 2020.

[22] Wall, G, & Mathieson, A., *Tourism: Change, Impacts and Opportunities*, Pearson Education: Harlow, 2006.

[23] Klarić, H. et al., *Studija prihvatnih kapaciteta turizma na području Splitsko-dalmatinske županije*, Institut za turizam: Zagreb, 2018.

SUSTAINABLE TOURISM AND INTANGIBLE CULTURAL HERITAGE: THE CIMBRIAN LANGUAGE IN LUSERNA/LUSÈRN AND GIAZZA, ITALY

SERENA LONARDI
Faculty of Strategic Management, Marketing and Tourism, University of Innsbruck, Austria

ABSTRACT

Intangible Cultural Heritage – i.e. practices, representations, expressions, knowledge and skills that a population recognizes as part of their cultural heritage – represents a distinctive factor for a tourism destination, as tourists are increasingly interested in unique experiences deeply connected with the authentic culture of the place. Intangible heritage is, therefore, part of the discourse on sustainability on tourism. Sustainable tourism activities, seen as activities that aim at not compromising the ability of future generations to meet their own needs and to access to natural and cultural resources, comprise intangible heritage that only survives and is available to future generations with continuous and daily practice by the population. One of the most important aspects of intangible heritage, according to The Convention for the Safeguarding of the Intangible Cultural Heritage, is the language since it stands for a unique worldview and is the vehicle of the collective memories and values. More than one-third of the languages spoken in the world today, however, are endangered. Previous studies proved that tourism strengthens the sense of cultural identity of minority groups and thus encourages the population to learn more, creating a virtuous circle of cultural revival. Nonetheless, current literature only partly considers the positive impacts of tourism on minority languages, while partially neglecting Italy, although it is home to 31 endangered languages. This study will fill in a gap by considering the case of Cimbrian people in Luserna/Lusérn and Giazza, Italy, where a population with Bavarian origins still speak the dialect of their ancestors. Through qualitative interviews, this study aims to analyse whether and to what extent tourism in these two small Italian towns can be seen as sustainable, i.e. whether it contributes to the revitalization of this particular language.

Keywords: sustainable tourism, minority language, Cimbrian, intangible cultural heritage, cultural tourism, qualitative methodology.

1 INTRODUCTION

Extensive research in the tourism field has proven that tourists are increasingly interested in unique and educational experiences that showcase an authentic culture of the place they are visiting [1]–[3]. As a consequence, cultural tourism has experienced a significant change: not only has it grown considerably, becoming the primary motivation for tourists [4], but it has also moved away from a more traditional view of tangible heritage, to include also intangible cultural heritage [5]. Intangible heritage has been defined by the UNESCO's Intangible Heritage Convention as "the practices, representations, expressions, knowledge, skills – as well as the instruments, objects and cultural spaces associated therewith – the communities, groups and, in some cases, individuals recognize as part of their cultural heritage" (p. 5), including traditional languages [6].

Local traditions and practices are particularly in danger of extinction because they need constant practice by the population to be transmitted from one generation to the next and, thus, survive [7]. For instance, more than one-third of the languages spoken in the world today are endangered [8]. Therefore, every activity that aims at preserving intangible heritage can be considered sustainable, as long as it pursues the availability of traditions and practices for future generations [9], [10]. One of these sustainable activities can be tourism [7].

WIT Transactions on Ecology and the Environment, Vol 248, © 2020 WIT Press
www.witpress.com, ISSN 1743-3541 (on-line)
doi:10.2495/ST200051

On the one hand, languages have exactly that connection with the cultural identity of a population, as they are the main vehicle of the collective memories and values and therefore stand for a unique worldview and knowledge [11], [12]. As a consequence, traditional minority languages have a huge potential application in tourism and represent a distinctive value and an asset for the destination [13]–[15]. On the other hand, it has also proven that tourism helps revitalize those languages since it becomes a matter of pride for the local communities, who are encouraged to learn more about their heritage [1], [16], [17].

Although being home to 31 endangered minority languages [6], Italy is still partially neglected by the literature. This study attempts to fill in this gap in the literature by analyzing, through qualitative research [18], which role tourism in Luserna/Lusérn (Trento) and Giazza (Verona) has in the ongoing revitalization process of the Cimbrian language.

This paper will be structured as follows: first, the relevant literature regarding the mutual relationship between minority languages and tourism will be summarized. Then, two case studies will be presented and the methodology used to collect the data will be explained. The third paragraph will show the main results. Finally, the results will be discussed and conclusions will be drawn.

2 LITERATURE REVIEW

By now it is common knowledge that the concept of sustainable development can be thought in terms of three spheres or dimensions, also called pillars: environment, economy and society [19]. The three pillars contain various important sub-aspects; for instance, some researchers argue that culture is an important aspect of the pillar of social sustainability, to the point that it should be considered a pillar on its own [20]. From the concept of sustainable development, many other concepts and definitions were inferred. Although there is still no official definition of sustainable tourism, an adapted definition of sustainable development from the Brundtland Report [19, p. 37], could read "(sustainable tourism is tourism) that meets the needs of the present without compromising the ability of future generations to meet their own needs". Not surprisingly, this is particularly true for intangible cultural heritage, as it will be available for future generations only with a continuous practice by the population. Therefore, sustainable tourism should also aim at preserving and revitalizing intangible heritage, including traditional minority languages, which if not spoken on an everyday basis, fall into oblivion [1]. The Convention for the Safeguarding of the Intangible Cultural Heritage, in fact, characterize languages as one of the most important aspects of intangible heritage, both as cultural artefacts per se and as fundamental for cultural transmission from one generation to the next [6].

According to the Council of Europe [21], minority languages are those languages spoken by a group of people that are not concentrated on a specific part of the State or by a group numerically smaller than the speakers of the official language. Traditional languages are "traditionally linked with an ethnoculture" [12] (p. 20). Since speakers are few, traditional minority languages are often threatened with extinction [22]. Language endangerment and language loss have been a matter of concern for linguists for decades due to reasons connected to a loss of cultural and intellectual diversity, linguistic interest, identity, history and human knowledge, i.e. information contained in a language [22], [23]. As each language stands for a unique worldview and is the vehicle for collective memories and values, losing it could then lead to terrible consequences for a people and its cultural identity, as "language shift generally and basically involves…quite devastating and profound cultural change" [12] (p. 16). Methods to work towards language preservation include language documentation, place names, education and mass-media. Not all linguists, however, agree on which method works best [23]. For instance, language commodification, the use of languages as commodities in fields such as advertising and tourism, is discussed [13], [14].

The relationship between minority languages and tourism has already been highlighted by the literature. Some researchers point at some controversies regarding authenticity and commodification, since, when languages are used in tourism they acquire different meanings than those they have in the community as they become marketable products [13], [14]. Another reason for concern is the fact that tourism brings the need for people to learn a global language, like English, in order to be able to interact with international visitors, which, according to some, competes with the process of language preservation [24]. However, as tourists seek authentic cultural experiences, traditional languages represent an asset for the destination [25].

Various authors provide a more skeptical insight on the issue. Languages can be used as a sign for authenticity and, as such, represent an asset for the destination; however, they thus become a marketable commodity, causing problems of commodification. The Irish language, for instance, is often used in combination with other symbolic resources, making it more understandable for tourists but undermining the authenticity of the language [26]. Some perplexities are also expressed by Kelly-Holmes and Pietikäinen, who considered how cultural museums represent language as an artefact and its impact on the population. Museums represent the Sami language in a folkloristic way and thus contribute to maintaining the ideologies and hierarchies. However, valorizing the language and demonstrating its value leads to a revitalisation of the Sami language and culture [27]. Burden also analyzed the problem of the representation of intangible heritage in museums, with an emphasis on the Afrikaans Language Museum. The display and presentation of intangible heritage are of course challenging, but this has some positive and important impacts: firstly, languages have to be established as a written language, which has not occurred yet; furthermore, the museum provides education for local people as well [28].

On the other hand, however, the process of cultural revival caused by ethnic tourism has been a matter of growing interest among the academic community [29]. Nowadays, many minority populations and Indigenous people are striving to (re)invent new strategies in order to survive as a distinct cultural group and make a living out of it. This economic space has been, for many Indigenous groups, heritage tourism. The deployment of the local language is an ideal marker of authenticity, thus highlighting the distinctiveness and remote otherness of the experience, which has become added attributes of touristic places. The traditional culture and language of the Sàmi, Indigenous people living in Northern Scandinavia and West Russia, was, for instance, an asset for the development of tourism in the region [30], [31]. This is valid for other communities as well: the production of arts and crafts as souvenirs by the Pataxo Indians (Brazil) generated a sense of group identity and helped legitimize the presence of the Indians in the region. Tourism also generated a revival of Indigenous names and language, as Indigenous people realized it is more appealing to present oneself with traditional names and to use the traditional language to brand restaurants and hotels [29]. In Haida Gwaii (British Columbia, Canada), Haida people decided to incorporate their traditional language into the management and development of touristic products. The benefits extended to the community itself: thanks to tourists' genuine interests in their traditional culture, Indigenous people started feeling proud about their origins and wanted to learn more [17]. In Mexico, tourism also appears to be an important factor that motivates Indigenous people and mestizos in Puebla to preserve and revitalize their languages, traditions, and culture, by providing them financial assistance [16], [32].

Minority communities in northern Italy were partially considered as well. In South Tyrol, the Ladin minority group offers the visitors an added value, with their traditions, customs, architecture and even languages [15]. Other projects considered the Cimbrian minorities in Giazza and Luserna/Lusérn and proved that the Cimbrian culture represents an added value for the tourists and an asset for the tourism industry [33] and, on the other hand, that tourism has a positive impact on the community [34].

The practical implications of this research are evident: aspects of minority culture, such as languages, represent a distinctive factor for the destination and therefore a competitive advantage [8], [15], [25]. Moreover, according to the studies presented here, tourism also has a sustainable impact on the local population, since it encourages language preservation [17], [25], [31].

3 CASE STUDIES AND METHODOLOGY

3.1 Case studies

The brief literature reported here shows that Italy has only been briefly touched upon, even though the UNESCO's Atlas of the World Languages in Danger identifies 31 endangered languages there [8]. This study will fill in a gap in the literature by considering the Cimbrian communities of Giazza and Luserna/Lusérn (Italy). Cimbrian people are descendants of Bavarian immigrants who, between the 11th and 16th centuries, were called to work woods and cultivate new terrains in the alpine mountains [35]. Due to the isolation of the mountain villages where they settled, they retained their language until the present day, an Upper Germanic variety called Cimbrian language [36].

This language was once spoken in three main territories of North-eastern Italy: the so-called *Thirteen Communities* (near Verona), *Seven Communities* (near Vicenza) and Luserna/Lusérn – Lavarone (near Trento). As the three areas were isolated from the city, they retained their traditional language, an Upper German variety derived from Southern Bavarian, called the Cimbrian language. Currently, this language is spoken only in three districts: Luserna/Lusérn in the Trentino region, just by some elders in Giazza near Verona and in Mezzaselva di Roana near Vicenza [37]. As a consequence, according to the UNESCO's Atlas of the World's Languages in Danger, the Cimbrian language is definitely endangered, i.e. children do not learn it as their mother tongue anymore, even though parents may speak it to them [8]. Nonetheless, a clarification here is needed, as there is a significant difference between the three communities above mentioned. Nowadays, the community where the majority of the population still speaks the language is Luserna/Lusérn, where according to the latest census in 2011, 238 out of 279 people living there spoke the language, including the children [38]. The situation is a lot more serious in Giazza, where, in 2001, only 19 people out of a population of 131 could speak the Cimbrian language and another 24 could understand it. Young people could neither speak it nor understand it [37]. However, the situation is worst in Roana, where only about 10 elderly people speak the Cimbrian language [37]. This study will only consider Giazza and Luserna/Lusérn.

Giazza is located in the Lessinia area, the Alp region near Verona. The area is visited by hikers, people who practice mountain sports and, in a limited number, by winter lovers [33]. Giazza, for instance, is the starting point for many hikes: from there you can reach the Foresta Demaniale di Giazza, but also the Valle del Fraselle and, the more challenging Gruppo del Carega (2259 m.a.s.l.) [39]. This little town lacks a tourism offer specifically focused on the Cimbrian language, even though visitors can find some general cultural tourism offerings. For instance, the Centro di Cultura Cimbra – Museo dei Cimbri Ljetzan – Giazza, is a museum that has a wide explanation about the history of the area and the peculiarities of the Cimbrian culture but is rarely open. There are also events connected to the Cimbrian culture, like the well-known Festa del Fuoco (*Waur Ljetzan* in the traditional language, which can be translated into 'fire feast'), held every year on the summer solstice, 23 June [39]. Another event that is becoming increasingly popular is the starting of the *carbonara* – charcoal pile – a traditional method to produce charcoal, particularly widespread in the region. This tradition was slowly dying until about 40 years ago, when the owner of the Osteria Ljetzan – a restaurant in the main square – realized that it would be much more convenient to produce their own charcoal and therefore, decide to restore this ancient tradition. In the last few years, thanks to the commitment of the restaurant owner, who continually promotes on social media and has even placed a live-webcam near the *carbonara*, this event has become more and more famous and important among the local population, and has slowly attracted visitors from outside the community as well [40].

Thanks to its peculiar characteristics, Luserna/Lusèrn is a perfect tourism destination for sport and mountain lovers, but also for those interested in history and culture. Not only is it nestled in the Alps, but is also a living witness of World War One, with remains of forts, trenches and the Cimbrian culture. The Provincia Autonoma di Trento is trying to promote the Cimbrian minority through education, social activities but also through tourism. Luserna/Lusérn is part of the DMO "Alpe Cimbra" (Cimbrian valley) that also includes the neighboring Folgaria and Lavarone. The name "Alpe Cimbra" is quite new, which proves the increasing interest in this minority culture. The Museum Centro di Documentazione Luserna and the Haus von Prükk have a goal to bring awareness about the Cimbrian minority and of preserving its culture and language [41]. The town also offers festivals, cultural events and hikes, all related to the Cimbrian culture [34]. A project carried out by Eurac specifically focuses on the Cimbrian culture of Luserna/Lusèrn, though only marginally considering the language. Results revealed that this minority culture represents an attractive distinctive factor for the destination and therefore encourages the population to make every effort to maintain and revitalize it [34].

The fact that the two minority communities in Giazza and Luserna/Lusérn have the same origin and used to speak the same language but are now so different, will allow for a fruitful comparison and analysis of the role of tourism in the maintenance and revitalization of the Cimbrian language. Table 1 provides a summary of the two case studies. This research will mainly examine to what extent tourism in Giazza and Luserna/Lusèrn can be considered sustainable, in the sense that it guarantees that the Cimbrian language and culture will continue to be available for future generations.

3.2 Methodology

Since this research deals with cultural and anthropological aspects of the population, it will use a qualitative methodology [18]. Qualitative methodologies consider few cases but have the considerable advantage to provide a holistic, in-depth and authentic understanding of the phenomenon being studied [42].

Until now, eleven in-depth interviews [18], [43] of Cimbrian speaking people working in tourism, as well as key players in both tourism and governance of the two towns, have been conducted. Participants were selected through snowball sampling. The interviews took place between summer 2018 and February 2019. Other interviews were planned, but the restrictions imposed by the Corona-crisis forced to an interruption of the research.

Therefore, in order to improve the generazability of the study, ten to fifteen subjects will be contacted in the future. The framework for the interviewees will be narrative, i.e. will have few questions (max. three) and will be more akin to an open conversation, where the participants will decide what they think is important and needs particular attention [42], [43]. This will allow for a collection of personal feelings and opinions about tourism and the use of particular aspects of the Cimbrian culture, like the language, in tourism activities. Interviews were face-to-face: participants were interviewed in a context familiar to them, i.e. either Giazza or Luserna/Lusérn and in their houses or in the workplace (museum or restaurant).

Questions were carefully selected from the literature review [43]. With written consent from the interviewees, interviews were recorded, and then transcribed by the author. Interview transcripts were then translated into English by the author and analyzed, coded and compared with the help of a computer software, MaxQDA [44].

Table 1: Case studies.

	Giazza	Luserna/Lusérn
Location	Verona	Trento
Speakers of the Cimbrian language left	19/131	238/279
Tourism	Mountain tourism destination	Mountain tourism destination, people interested in history
	Mainly summer	Summer and winter
	Museo dei Cimbri Ljetzan – Giazza	Museum Centro di Documentazione Luserna and Haus von Prükk
Traditions retained until today	*Waur Ljetzan* and *carbonara*	Bobbin lace and *vorprennan in martz*

4 RESULTS

This study represents a steppingstone in the analysis of the positive impacts of tourism on minority languages, especially in a partially neglected context like Italy. The analysis of two Cimbrian communities will contribute to fill in a gap in the literature in that it considers minority languages as sustainable tourism resources. This paragraph will present the main and more significant results of this study.

All interviewees were aware that their traditional language is seriously threatened and therefore, the process of language preservation is urgent and of the utmost importance. However, aligned with the literature, a significant difference was found between Luserna/Lusérn and Giazza. Participants in Giazza were almost resigned that the language is disappearing, whereas in Luserna/Lusérn the interviewees were much more positive.

> "From what I can see, the language is slowly dying. The accent has changed…the children are not learning it" (restaurant owner in Giazza).
> "I'd say 80% of the population here can understand the language" (staff member at the Centro Documentazione Luserna).

A more similar view between the two locations was found in the awareness that the Cimbrian culture and language represents an asset for the destination:

> "The community now is proud of the language [...]. There is a growing interest from outside, as well. When I'm attending the conferences in Verona, people are very keen to ask for clarifications [...]. The same when I take visitors on the hikes around Giazza" (Cimbrian speaker, expert and writer).
> "Some restaurant owners translated their menus in Cimbrian, because it is an added value. People who come here and discover that there is a minority culture and language, appreciate it and want to know more" (staff member at the Centro Documentazione Luserna).

As a consequence, the Cimbrian language is part of the tourism experience offered to visitors. In Giazza, the language is part of the landscape since all signs and plaques are also translated in Cimbrian. For instance, along the little hike leading to the *carbonara* visitors can find signs describing this tradition, which are translated into German, English and Cimbrian. Moreover, in the museum, various panels explain the history and origin of the language together with the meaning of some words. The Cimbrian language is integrated into both the Centro Documentazione di Luserna and the museum Haus von Prükk: panels in both museums are in four languages (Italian, German, English and Cimbrian), greetings from the staff is often in Cimbrian and a specific section of the Centro Documentazione di Luserna is dedicated to the origin and evolution of the language, as in Giazza.

Other aspects of the culture contribute to differentiate the destination. In Luserna/Lusérn one of the hikes visitors can find is thematic and evolves around traditional Cimbrian legends and creatures of the local folklore. This, according to the interviewees, represents an added value:

> "We offer a hike that can be found only here in Luserna; it's called nå in tritt von Sambinélo, which means 'on the trail of the Slavanello'. It presents the Cimbrian legends, which adds value to the hike...It the most popular hike here, visited by thousands of people every year" (staff member at the Centro Documentazione Luserna).

As already mentioned in the presentation of the case studies, the tradition of the *carbonara* has become increasingly well-known in the Lessinia area around Giazza: the event is attended by dozens of people and provides the Osteria Ljetzan with a "story to tell" since the restaurant now uses the coal produced with this method to cook the local trout. As an appreciation for its local cuisine and restoration of this tradition, the Osteria Ljetzan is an official Lessinia Park branded trader.

> "It's an added value. No one cooks with the coal he produces by himself. You bet I'm so proud of that!" (owner of Osteria Ljetzan in Giazza).

Other traditional practices have been revitalized in Luserna/Lusérn as well, thanks to the growing interests by people from outside the community. For instance, "*bruciar el martzo/vorprennan in martz*" (burning March), a traditional event in the Cimbrian tradition in Luserna/Lusérn, which consists in lightening a bonfire to welcome spring, was slowly disappearing but is now attended by hundreds of families from North-eastern Italy. Another example in Luserna/Lusérn is the tradition of the bobbin lace (*tombolo*), widespread in the Cimbrian town for centuries, has been growing despite the modernization and has made the town renowned among the enthusiasts around Italy and even abroad.

The integration of particular aspects of this minority culture in various tourism experiences led to satisfaction and loyalty since many interviewees felt that many visitors in both Giazza and Luserna/Lusérn are or will become returning visitors. They felt this loyalty to the destination could be ascribed to Cimbrian words and traditions, rather than the natural landscape, especially in Luserna/Lusérn.

A significant art of the conversations was dedicated to the topic of sustainability, which is of primary importance when dealing with delicate aspects of minority cultures. In this case, the concept of sustainability evolves around culture preservation, especially as regarding the endangered Cimbrian language. In Giazza, courses of Cimbrian for adults take place almost every year, but, unfortunately, rarely the schools in the area include the Cimbrian language as an optional class. Moreover, more and more experts are visiting the town and documenting the language. However, all interviewees were skeptical that this could lead to a real language preservation in the long term, i.e. language will be transmitted to future generations, even though visitors show interest and want to know more. On the contrary, participants in Luserna/Lusérn were certain that the tourism field in the town contributes to the process of language preservation. Since it is a small and isolated town, it does not offer many job opportunities to its inhabitants. Thanks to this economic activity, many people can remain in the town and work there in companies that are somehow related to the tourism field. All interviewees highlighted that besides the museums and the Kultinstitut Lusérn, nine restaurants operate in Luserna/Lusérn, which is a huge number for such a small town:

> "So, if there's the opportunity to keep these companies here in Luserna, even the language will benefit from that. All these families stay here and will keep speaking the language and pass it on to their children" (staff member at the Centro Documentazione Luserna).
>
> "We have the chance to preserve and give value to the language, we can use it in the official documents, it is taught in schools, the news is in Cimbrian. The problem is the economic development. […] Only 200 people actually stay in Luserna during the week, we are at the threshold of survival for a community" (former mayor of Luserna/Lusérn).

As a consequence, people in Luserna/Lusérn stressed the importance of the tourism field for the town and the survival of the minority culture. They also highlighted that is consequently extremely important to continue giving value to the Cimbrian culture and language, in order for the destination to maintain this uniqueness, that is a source for the attractiveness of the destination.

The concept of sustainability is also related to a genuine and authentic display of the local culture. The local population in Luserna/Lusérn, in fact, is committed to present the tourist a culture, as authentic as possible:

> "I think in the Alpe Cimbra the community is committed to a genuine preservation and promotion of the culture and that is fine […]. It is ok to re-interpret some traditions, even in Lavarone and Folgaria, but we should not distort them" (director of the Centro di Documentazione Luserna and cultural liaison of the Kulturinstitut Lusérn).

The same concern about an authentic presentation of the Cimbrian culture was also expressed by the interviewees in Giazza, with two different approaches. On the one hand, tourists are genuinely interested in the Cimbrian culture and language. On the other hand, those who are not familiar with the language do not feel comfortable in using and promoting it.

> "Overall, this growing interest in the Cimbrian culture is genuine. More people in the region are getting to know this culture and appreciate it" (Cimbrian speaker, expert and writer).
> "I can't promote the Cimbrian culture and language if I don't speak it myself" (restaurant owner of the osteria in Giazza).

Obviously, some tensions have arisen from the commodification of the Cimbrian culture. For instance, in Giazza, visitors of the Lessinia Park can find various "Cimbrian" products, like Cimbrian cheese, Cimbrian pizza and bread, Cimbrian strudel etc. Even though it is true that Cimbrian people had a centuries-long tradition in cheese production, all the other products did not previously exist.

Another example is the "Cimbrian ring", sold in Folgaria, the ski-destination near Luserna/Lusérn. The guide of the museum explained that she did not know anything about such a piece of jewelry in her traditional culture and she also did not know what was meant with this souvenir. Even though at the beginning she was surprised and rather disappointed that such a souvenir was sold without consulting a committee of experts on the Cimbrian culture and language, she also argued that this commodification is ok until it generates curiosity in the visitor and becomes an opportunity to visit Luserna and learn more about the Cimbrian culture. She, as well as the director of the Centro Documentazione Luserna, present another example of commodification: "*bruciar el martzo/vorprennan in martz*", which, in the last decades, has taken place on the last Saturday afternoon of February rather than the last day of the month as it used to be because in this way more people can participate and contribute to keeping this tradition alive.

To conclude, all interviewees believed more has still to be done in order to give value to this minority community through tourism activities. Not surprisingly, one of the most iconic examples mentioned was the possibility to expand the use of the language, especially in the signs of public spaces: streets, plaques, but also restaurants and bars:

> "I think tourism contributes to language preservation, but we could definitely do more…The bar next to the town hall for us is not the bar ristorante Rossi', it's the platzbirt (bar in the square). Further on, there's the old Andreashofer, we don't call it bar Ferdi as the sign says. We should restore those traditional names. I firmly believe we should give more visibility to the language"
>
> (director of the Centro di Documentazione Luserna and cultural liaison of the Kulturinstitut Lusérn).

5 DISCUSSION

The analysis of the results also showed that tourism indirectly contributes to the preservation of minority languages, especially in Luserna/Lusérn. Differently from the literature, however, this process of preservation was activated not so much because of an enhanced sense of pride among the population [16], [17], [25] – the interviews in Luserna/Lusérn showed that this sense of pride was already present among Cimbrian people – but because tourism created various job opportunities and, therefore, allowed for many families to stay in the town and maintain the relations with other inhabitants and Cimbrian speakers, fundamental for the maintenance of the language [12]. This confirms what was theorized in the Brundtland Report, i.e. the fact that the three pillars economy, society and environment of the sustainable development are inherently related to one other

[19]. Social, and particularly in this case, cultural sustainability is inevitably linked with the pillar of economic sustainability. Only with a sustainable economic development, that guarantees that people in Luserna/Lusérn can find a job in the town or in neighboring destinations and are not obliged to move, the language will survive [19], [29]. The Cimbrian culture, and especially the traditional language, is fundamental for this process of sustainable economic development, since it becomes an asset to differentiate the destination and, thus, a pull factor for visitors [15], [25], [27], [33], [34].

The study also confirmed that there are significant differences between the two communities analyzed. On one hand, in Luserna/Lusérn the Cimbrian language is still spoken by the majority of the population and transgenerational transmission is somehow guaranteed, also thanks to childcare programs in Cimbrian and Cimbrian classes taught at schools in the neighborhood. On the other hand, in Giazza there is no solid and official program that aims at maintaining the language: this relies just upon the (voluntary) work of single persons and researchers, who document the language and offer courses for a group of 25/30 adults per year [36], [37]. As a consequence, the use of the language in tourism is more frequent in Luserna/Lusérn than in Giazza, which leads to the conclusion that tourism cannot be seen as the only tool to achieve a proper language preservation. Tourism only contributes to a complete and effective language preservation if the language is only vulnerable or definitely endangered. The results showed the hesitation of participants in Giazza to talk about a language transmission to future generations, which is at the basis of a proper language preservation [12].

However, even though this study proves that the tourism field does not always lead to a complete and durable preservation of minority languages, it proved that it certainly contributes to the revitalization and preservation of other aspects of the intangible cultural heritage of a minority population, like traditional practices as the *carbonara* in Giazza, as well as the *vorprennan in martz* and the bobbin lace in Luserna/Lusérn [7].

Though the more frequent use of the language and traditional practices was definitely somehow related to tourism, we cannot talk about a process of pure commodification. This was not exclusively aimed at tourists' satisfaction and did not modify the socio-cultural balance of the Cimbrian population. People were sincerely driven by a sense of pride and genuinely wanted to present and show their language and their culture to the visitors.

6 CONCLUSIONS

This explorative study brings awareness not only on the value of minority languages for the tourism field but especially on the value of tourism for intangible cultural heritage, and especially minority languages. Aligned with the literature, the results presented in the previous paragraph confirmed that minority cultural group, with their traditional languages, represent a differentiating factor for the destination and, thus, contribute to attracting visitors [15], [25], [27], [33], [34].

Results also showed how important this is for an isolated minority community, since it allows for an economic sustainability of the town, i.e. creation of job opportunities that contributes to avoiding the depopulation of the town. Economic sustainability is intrinsically related to social/cultural sustainability because it guarantees language transmission and, therefore, language preservation [19]. As explained earlier, however, the tourism field seems to contribute to language maintenance only in those communities where the language is still widespread among its inhabitants, even though interviewees said that more could be done to integrate the language in the tourism activities and differentiate the destination. Nonetheless, tourism definitely contributes to the revitalization and preservation of other intangible traditional practices, as no significant differences were found between Giazza and Luserna/Lusérn. Therefore, the tourism industry can definitely be considered a sustainable activity [19].

This study hopes to represent a stepping-stone in the analysis of sustainable tourism and minority language revitalization, especially in an Italian context. Further research is needed, in order to analyse more in-deep the case study presented here, but also include other minority communities in the discourse. Finally, considering the situation the world is living right now, with the global pandemic caused by Covid-19 and the consequent lockdown of most countries worldwide, more research should be dedicated to domestic and proximity tourism. Further research should be dedicated to understand how much tourists are now interested in (re)discover the "hidden gems" in their neighbourhood, like the Cimbrian communities of Giazza and Luserna/Lusérn.

REFERENCES

[1] Kim, J., Ritchie, J.R.B. & McCormick, B., Development of a scale to measure memorable tourism experiences. *Journal of Travel Research*, **51**(1), pp. 12–25, 2012.

[2] Kolar, T. & Zabkar, V., A consumer-based model of authenticity: An oxymoron or the foundation of cultural heritage marketing? *Tourism Management*, **31**(5), pp. 652–664, 2010.

[3] van den Perge, P. & Keyes, C.F., Introduction. Tourism and re-created ethnicity. *Annals of Tourism Research*, **11**, pp. 343–352, 1984.

[4] UNWTO, Report on Tourism and Culture Synergies, Madrid: UNWTO, 2018.

[5] Richards, G., Cultural tourism: A review of recent research and trends. *Journal of Hospitality and Tourism Management*, **36**, pp. 12–21, 2018.

[6] UNESCO, Convention for the safeguarding of the Intangible Cultural Heritage, 2018. https://ich.unesco.org/en/convention. Accessed on: 21 Aug. 2019.

[7] Kim, S., Whitford, M. & Arcodia, C., Development of intangible cultural heritage as a sustainable tourism resource: The intangible cultural heritage practitioners' perspectives. *Journal of Heritage Tourism*, **153**(3), 1–14, 2019. https://doi.org/10.1080/1743873X.2018.1561703.

[8] Moseley, C. (ed.), *Atlas of the World's Languages in Danger*, 2010. www.unesco.org/languages-atlas/en/atlasmap.html. Accessed on: 26 Aug. 2019.

[9] Chhabra, D., Proposing a sustainable marketing framework for heritage tourism. *Journal of Sustainable Tourism*, **17**(3), pp. 303–320, 2009.

[10] Chhabra, D., Healy, R. & Sills, E., Staged authenticity and heritage tourism. *Annals of Tourism Research*, **30**(3), pp. 702–719, 2003.

[11] Krauss, M., The world's languages in crisis. *Language*, **68**(1), pp. 4–10, 1992.

[12] Fishman, J.A., *Reversing Language Shift, Theoretical and Empirical Foundations of Assistance to Threatened Languages*, Multilingual Matters: Clevedon, 1991.

[13] Kelly-Holmes, H., *Advertising as Multilingual Communication*, Palgrave Macmillan: Basingstoke, 2006.

[14] Heller, M., Pujolar, J. & Duchêne, A., Linguistic commodification in tourism. *Journal of Sociolinguistics*, **18**(4), pp. 539–566, 2014.

[15] Pechlaner, H., Lange, S. & Raich, F., Enhancing tourism destinations through promoting the variety and uniqueness of attractions offered by minority populations: An exploratory study towards a new research field. *Tourism Review*, **66**(4), pp. 54–64, 2011.

[16] Greathouse-Amador, L.M., Tourism: A facilitator of social awareness in an indigenous Mexican community? *Review of Policy Research*, **22**(5), pp. 709–720, 2005.

[17] Whitney-Squire, K., Sustaining local language relationships through indigenous community-based tourism initiatives. *Journal of Sustainable Tourism*, **24**(8–9), pp. 1156–1176, 2016.

[18] Richards, G. & Munsters, W., *Cultural Tourism Research Methods*, CABI: Wallingford, 2010.

[19] World Commission on Environment and Development, *Our Common Future*, Oxford University Press: Oxford, 1987.

[20] Soini, K. & Birkeland, I., Exploring the scientific discourse on cultural sustainability. *Geoforum*, **51**, pp. 213–223, 2014.

[21] Council of Europe, CETS 148 – European Charter for Regional or Minority Languages. *European Treaty Series*, 1992.

[22] Crystal, D., *Language Death*, Ernst Klett Sprachen, 2000.

[23] Bradley, D. & Bradley, M., *Language Endangerment*, Cambridge University Press: Cambridge, 2019.

[24] Bruyèl-Olmedo, A. & Juan-Garau, M., Minority languages in the linguistic landscape of tourism: the case of Catalan in Mallorca. *Journal of Multilingual and Multicultural Development*, **36**(6), pp. 598–619, 2015.

[25] Lonardi, S., Martini, U. & Hull, J.S., Minority languages as sustainable tourism resources: From Indigenous groups in British Columbia (Canada) to Cimbrian people in Giazza (Italy). *Annals of Tourism Research*, **102859**, 2020.

[26] Moriarty, M., Indexing authenticity: The linguistic landscape of an Irish tourist town. *International Journal of the Sociology of Language*, **232**, pp. 195–214, 2015.

[27] Kelly-Holmes, H. & Pietikäinen, S., Language: A Challenging resource in a museum of Sámi Culture. *Scandinavian Journal of Hospitality and Tourism*, **16**(1), pp. 24–41, 2016.

[28] Burden M., Museums and the intangible heritage: The case study of the Afrikaans Language Museum. *International Journal of Intangible Heritage*, **2**, pp. 82–91, 2007.

[29] de Azaredo Grünewald, R., Tourism and cultural revival. *Annals of Tourism Research*, **29**(4), pp. 1004–1021, 2002.

[30] Kelly-Holmes, H. & Pietikäinen, S., Commodifying Sámi culture in an indigenous tourism site. *Journal of Sociolinguistics*, **18**(4), pp. 518–538, 2014. DOI: 10.1111/josl.12092.

[31] Pietikäinen, S., Kelly-Holmes, H. & Peer, H., The local political economy of languages in a Sámi tourism destination authenticity and mobility in the labelling of souvenirs. *Journal of Sociolinguistics*, **15**(3), pp. 323–346, 2011. DOI: 10.1111/j.1467-9841.2011.00489.x.

[32] Greathouse-Amador, L.M., Tourism and policy in preserving minority languages and culture: The Cuetzalan experience. *Review of Policy Research*, **22**(1), pp. 49–58, 2005.

[33] Ugolini, M. & Costa, K., I cimbri come fattore di attrattiva turistica? *Quaderni di lingue e letterature*, **34**, pp. 113–133, 2009.

[34] Dal Bò, G., Tomaselli, A., Chiochetti, E., Ralli, N. & Stanizzi, I., *Turismo e minoranze a Luserna/Lusérn,* EURAC Research: Bolzano, 2012.

[35] Baum, W., *Geschichte der Zimbern: Gründung, Sprache und Entwicklung der südbairischen Siedlungen in den VII und XIII Gemeinden in Oberitalien = Storia dei Cimbri*, Curatorium Cimbricum Bavarense (Storia dei Cimbri): Landshut, 1983.

[36] Bidese, E., Italy I: Cimbrian. *Varieties of German Worldwide*, eds H. Boas, A. Deumert & L.M. Louden, P. Maitz, Oxford University Press: Oxford (in press).

[37] Stringher, A., *Censimento dei parlanti Cimbro nell'isola linguistica di Giazza: Consistenza della parlata tedesca dei tredici comuni veronesi dal XVII al XX secolo*, Comune di Selva di Progno: Selva di Progno (VR), 2012.

[38] Servizio Statistica della Provincia Autonoma di Trento, *Rilevazione sulla consistenza e la dislocazione territoriale degli appartenenti alle popolazioni di lingua ladina, mòchena e cimbra* (15° Censimento generale della popolazione e delle abitazioni – dati definitivi)*, Trento, 2014.

[39] Alta Lessinia, *I Cimbri e Giazza*. www.altalessinia.com/cimbri-giazza-lessinia/. Accessed on: 14 May 2020.

[40] Alta Lessinia, Osteria Ljetzan. www.altalessinia.com/osteria-ljetzan/. Accessed on: 14 May 2020.

[41] Lusern, www.lusern.it. Accessed on: 28 Aug. 2019.

[42] Denzin, N.K. & Lincoln, Y.S. (eds), *The Sage Handbook of Qualitative Research*, SAGE: Los Angeles, 2011.

[43] Jennings, G.R., Interviewing: A focus on qualitative techniques. *Tourism Research Methods: Integrating Theory With Practice*, eds B. Ritchie, P. Burns & C. Palmer, pp. 99–117, CAB Publishing: Wallingford, 2004.

[44] Saldaña, J., *The Coding Manual for Qualitative Researchers*, SAGE: Los Angeles, 2009.

REVIEW OF SHORT-TERM RENTALS AND THEIR POTENTIAL SOCIAL IMPACT

SELINA ANGEL & SEDEF DOGANER
Department of Architecture, University of Texas at San Antonio, USA

ABSTRACT
This paper reviews the development of short-term rentals (STRs) and discusses their potential social impacts within historic neighborhoods. The paper focuses on cultural heritage tourism, the characteristic of cultural travelers, and emphasizes the importance of community. The growth management of STRs in neighborhoods are also discussed in order to sustain tourism, but prevent displacement. South San Antonio neighborhoods along the San Antonio Missions World Heritage Buffer Zone are selected as a case study. These neighborhoods are originally outside of the tourist bubble but this fact has recently started to change because of the 2015 WHS inscription. The methodology includes review of short-term rentals in the San Antonio neighborhoods and data collection from the Development Services Department on regulated, unregulated, and grandfathered STRs. The findings of this paper will help cities understand the importance of managing STRs and its social impact, and how regulations can help keep the original essence of life in these historic neighborhoods and prevent gentrification.
Keywords: cultural heritage tourism, short-term rentals, social impact, historic neighborhoods, gentrification, heritage management, cultural sustainability.

1 INTRODUCTION

Short-term rentals (STRs), a residential property that is rented out for a short stay, can be seen popping up everywhere! Some would say that STRs are contributing to a “shared economy”, a peer-to-peer distribution of goods and services through community-based exchange. Others would argue that STRs are dramatically changing the social fabric of a community. As new policies are put into place to regulate the STR trend, sustainable development must be the focus of this new practice.

In 2015 San Antonio became the first city in Texas to become a United Nations Educational, Scientific, Cultural Organization (UNESCO) World Heritage Site (WHS). The designation includes five Spanish colonial missions: Mission Espada, Mission San Juan, Mission San Jose, Mission Concepción, and Mission San Antonio de Valero, more commonly known as the Alamo. A World Heritage Buffer Zone, approximately 5,755 acres of surrounding lands, was created to allow, not only for the environmental protection around the San Antonio missions, but also to moderate the impact of future development of the City of San Antonio [1].

The globalization of San Antonio’s WHS has caused an increase in visitation, tourist spending and new development. According to a 2018 National Park Service Report [2], there were 1,208,104 total recreation visits that contributed to $87,957 in visitor spending. Short-term rentals have not only intensified near the most popular tourist mission and attraction, The Alamo, but have spread around the other San Antonio missions. The impact on the neighborhoods surrounding the Southern missions may not be as dramatic as the neighborhoods surrounding the Alamo, due to location, but as the level of tourism increases, along with the rise of future development, the amount of STRs will also potentially grow. The paper discusses the potential social impacts of STRs, and analyzes the state of STRs in San Antonio in order to manage its growth.

www.witpress.com, ISSN 1743-3541 (on-line)
doi:10.2495/ST200061

2 CULTURAL HERITAGE TOURISM

Tourism is one of the economic generators of the industry. As of 2019, over 10% of the global GDP comes from travel and tourism, creating more than 300 million jobs worldwide [3]. In the United States alone the tourism industry provides jobs to 15.8 million people, generates an economic output of $2.6 trillion and $1.1 trillion in traveler spending [4]. There are many types of reasons why people travel, for leisure, business, religion, family, health, sports, education, or entrainment. However, 40% of travelers travel for cultural purposes [3]. These types of tourists are called "cultural tourists". Studies have shown that cultural tourists tend to spend more per day (an average of 60% more than the traditional traveler that travels domestically), stay longer, take more trips annually, are more likely to shop, tend to have higher household incomes, and are most likely to have a post-high school education. Even though many cultural tourists are baby boomers, there has been a growth in the millennial demographic in this market. 73% of millennials prefer to travel to a destination that is filled with arts and culture, while two-thirds of millennials desire an authentic experience when travelling [5]. In 2018, there was an estimated 47% of international arrivals of cultural tourists compared to a 16% of motivation-based tourists. The cultural tourist is "more interested in local customs, traditions, history, and culture of a place. It is no wonder that 84% of Member States of the World Tourism Organization (UNWTO) put emphasis in marketing and promoting cultural tourism" [6].

Cultural heritage tourism has numerous economic benefits such as the creation of new jobs and businesses. This stimulates the economy both directly and indirectly, causing a multiplier effect [7]. However, benefits of cultural heritage tourism extend beyond economics. The preservation of heritage within a community can lead to an enrichment of society by giving the place an identity. By honoring the roots and origins of a place, a link with the past is formed, allowing the modern-day residents to understand the present [8]. Heritage sites provide "real-world classrooms where both children and adults – local and visitors – can learn about the history, nature and/or culture of an area. Beyond these opportunities for education regarding the site or wider vicinity, lies the additional potential for increased awareness and understanding of resource preservation" [9].

For heritage tourism to exist, grow and prosper, there must be a cultural heritage within a place. However, the increase in the tourism economy can impose negative effects on a region. Displacement can occur when businesses and the real estate market take over tourist markets. Heritage tourism can disrupt local inhabitants and their quality of life, leading to gentrification, and the loss of identity [8]. This is the reason why there must be a balance between the promotion and protection of a place.

Over time, the accommodation industry has evolved to keep up with the different types of tourism and the changing needs of society. Hotels are no longer alone in the business. The emergence of short-term rentals (STRs) has almost taken over.

3 SHORT-TERM RENTALS

The earliest record of STRs can be traced back to the mid-1600s. They were known as vacation homes. The first property: Louis XIV's Palace of Versailles. It was built as a hunting lodge and became a destination for entertainment to political members of its era [10]. By the 1800s the first timeshares came to be in Europe. Families purchased homes to share throughout the year for vacationing, visiting at different times of the seasons. The invention of the telegraph in 1837 made it easier for renters to request a vacation home. By the 1950s the United States began to list vacation rentals in local newspapers [11]. Vacation rentals became hugely popular in the United States that the creation of the Vacation Rental Management Association (VRMA) was formed in 1985 in order to provide homeowners services in housekeeping, reservations, maintenance, marketing, accounting and other services [12]. As the internet was discovered, VBRO (Vacation Rentals by Owner) was the first online platform, launching in 1995. An online marketplace was born. Booking.com emerged one-year later, comparing hotel rates and Craigslist began ads for long- and short-term rentals around the same time [13]. In 2005, HomeAway purchased VBRO and five different online platforms, merging all inventories into one. But, it was not until 2008 that Airbnb stepped into the race to become one of the world's largest online marketplaces [11]. Within one year of its launch, Airbnb had 2,500 listings and approximately 10,000 members [14]. As of February 2020, Airbnb has 6 million plus listings across the globe, and 150 million plus users. According to iProperty Management, over 2 million people stay in an Airbnb per night [15].

Although, as stated above, there are many other STR platforms, Airbnb will be the focus of this study, due to its popularity, rapid growth, and success in the STR market.

3.1 Short-term rentals versus hotel chains

Before the emergence of STRs, staying with family or staying in a hotel was the traditional way to travel and explore places. These are still options today. Reasons why people prefer a hotel over a STR include the various amenities that come with a hotel stay such as room service, concierge services, shuttle transportation, location, kid friendly, business friendly, security, rewards and loyalty programs, brand recognition, free breakfast, a pampered experience and quick, attentive services that are a phone call away. As of 2019, there are over 700,000 hotels and 4 million plus hotel rooms worldwide [16]. Even with the addition of STRs in the market, hotels are still leading the accommodation industry, but that could soon change. According to Fig. 1 from Second Measure, compared to corporate hotel chains, Airbnb is experiencing faster growth at a high rate [17].

STRs, like Airbnb, are a global phenomenon and a market that grows practically every single day. Guttentag [18], the Director of the Department's Office of Tourism Analysis and an Assistant Professor in Hospitality and Tourism Management at the College of Charleston,

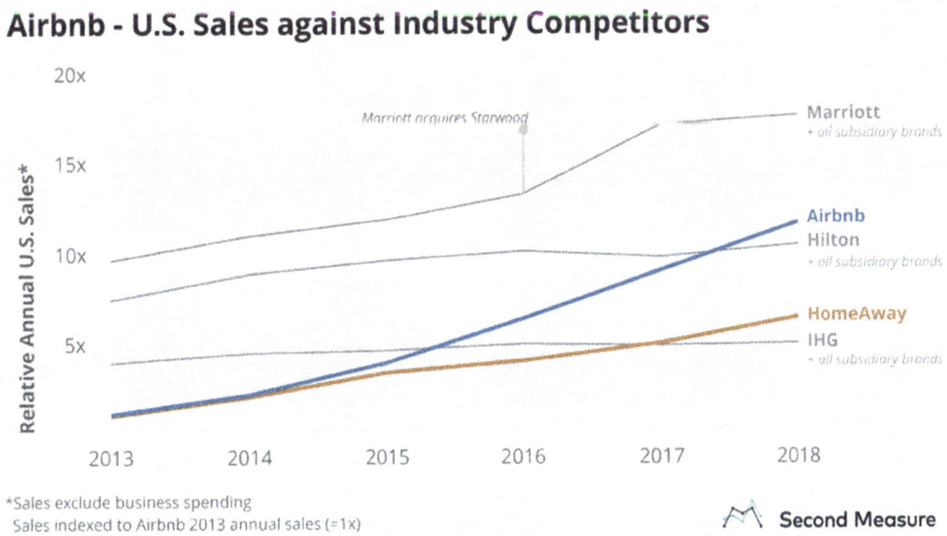

Figure 1: Airbnb: U.S. sales against industry competitors [17].

stated: "This rise is examined through the lens of disruptive innovation theory, which describes how products that lack in traditionally favored attributes but offer alternative benefits can, over time, transform a market and capture mainstream consumers".

What do hotels lack that STRs provide? It is not really a lack of attributes but what the consumer is willing to sacrifice. STRs are usually cheaper than a hotel because hotel amenities are not provided. There is no uniformity, predictability, or consistency in STRs, therefore there is no guarantee if a STR will live up to photos or descriptions that are provided through the online platform. Customer service may be slow or non-existent. Before recent laws, STRs were not regulated and many were against building and fire codes. Some STRs are "shared" spaces, where a bathroom, living room and kitchen may be shared with the person that resides in the house or other guests. In a "shared" space, the bedroom is the only space that is considered "personal" space but may lack a lock on the bedroom door. The location is not often central to places of attraction, but in nearby residential areas. Transportation may be necessary to get around but is not provided by the owner of the STR. Despite these downfalls, STRs are still on the rise. The alternative benefits of STRs strongly outweigh the sacrifices made.

According to iProperty Management, approximately $450 million in direct revenue is lost from the hotel industry to Airbnb alone [15]. Although, most of the time, it may be cheaper to stay in a STR, this revolution of modern-day travel comes with a price.

3.2 Short-term rentals and their social impacts

People who choose STRs over hotels often do because of the local experience. In fact, Airbnb's CEO and co-founder, Brian Chesky, when interviewed on multiple occasions, stated numerous times that "the number one reason people [choose] to travel on Airbnb is [that] they want to live like a local" [19]. The new tagline from Airbnb, "Belong Anywhere" has reflected the creation of new markets of real estate in areas that are outside the tourist bubble, settling in well-established neighborhoods.

Figure 2: Airbnb's icon development graphic [19].

The consequences of STRs rapid increase outside tourist locations has recently been recognized. There is a lack of restriction. There is a lack of care. There is a lack of sustainability. Many would say that STRs have revitalized neighborhoods while others argue that STRs are leading to the displacement of long-term residents. As the demand for tourism continues to rise, new destinations will emerge, and travel will expand.

Regarding STRs, there are three major impacts STRs have on communities when it comes to achieving sustainability: economic, environmental, and social. This paper will only discuss the social impact of STRs.

3.2.1 Social impact

Travelers want an authentic insight of the city they travel to. STRs provide that with shared accommodation that is embedded within a community. Guests are immersed in the local context. 86% of guests who used Airbnb's have stated that the location was convenient and 77% of Airbnb guests chose the STR because they wanted to live like a local [20]. Travelers look to explore areas outside the beaten path and interact with locals to understand and educate themselves with the culture, historic heritage, everyday lifestyles, and traditions of a city. They want to know the secrets and the hidden gems of an area. They want to experience a meaningful connection to the city.

> "Authenticity is anchored in society. Normal daily life cannot be considered authentic because it is not unique (every day the same…), but there is an argument that asserts that everything a tourist experiences as authentic must be considered as such. The discussion does not stop there. One has to realize all the time that what may be authentic in the eyes of a tourist may be daily routine for a local. The opposite holds true, too: things that local people feel are special and authentic in their society may be ignored completely by tourists" [21].

The rise of STRs in residential districts is taking away the sense of a neighborhood. Houses remain vacant until the tourists arrive for the weekend. Neighbors are now strangers and change every week. A study showed that out of 6,508 Airbnb's in New Orleans 85% were owned by San Francisco and New York City investors. There is a loss of autonomy in the area and a loss of identity. Communities are becoming commodities. Everyday life becomes difficult for residents because the neighborhood becomes a theme park for tourists or on the other side of the spectrum ghost towns. STRs take "away the soul of [a] place and the social tissue that make[s the] place alive" [22]. There becomes a dislike for tourists in the neighborhood. There is frustration due to the lack of regulation and implementation from local authorities. People start moving away. Without these long-term residents, there is no true neighborhood. There is no longer authenticity.

3.2.2 Gentrification

STRs fuel gentrification and displacement in four ways described in a report by the Jane Place Neighborhood Sustainability Initiative [23]:

- With the prioritization of tourists over residents.
- The oversaturation of STRs in residential neighborhoods.
- Single operators with multiple listings.
- Spillover effects from the monetization of residential housing inflating overall costs for renters and homeowners.

When tourism generates a wealthy economy, public, private, and non-profit entities seem to focus their attention on tourists rather than residents. There is a priority for projects in tourist districts to beautify the area to be more presentable to travelers. This attracts developers to invest in the area. The same thing happens when STRs take over a neighborhood. Investors buy into the neighborhood. Even existing landlords will throw out long time renters to accommodate short-term renters because replacing long-term tenants with short-term visitors is more profitable. In Austin, Texas homeowners are charging almost six times the price of the local rent for an Airbnb and travelers are paying for it [24].

It is because of this "authentic atmosphere" tourists pay the price. The authenticity of the neighborhood was first driven by the lack of tourists in the area. The "edgy" part of town which are the "transitional neighborhoods or zones where people are on the margins of urban society", areas where the natural habitat of locals reside [25]. These places are often of a low income and minority demographic; the reason why it is so easy for investors to buy into the cheap real estate market that is sparking new urban tourism. "Compared to the traditional rental market, short-term rentals offer a number of benefits that enhance market efficiency for property owners, making them increasingly attractive for both local and global investors" [26].

These investors usually do not care about the authenticity of a place, yet alone the people that reside in the neighborhood. In fact, less than 20% of Airbnb's are true home-sharing residencies, where the homeowner is present [27]. That means that more than 80% of Airbnb hosts are not associated with the community in which they rent out and most likely have the STR as a side business. In return, this potentially causes problems for locals to find affordable long-term housing. "More houses, apartments and condos available on home-sharing platforms translates into fewer options for those who live [in the area] full time" [28]. The supply in the traditional private market is becoming less available, "provoking exclusionary displacement as new residents are unable to move in. The result is a tacit change in land use from residential to tourist uses that residents experience as a process of social injustice" [26].

4 ANALYSIS SHORT-TERM RENTALS IN SAN ANTONIO, TX

In 2018 San Antonio was listed as one of the world's 50 best places to travel in the Travel+Leisure magazine, making it only one of three U.S. cities [29]. San Antonio brings in about 37 million visitors a year, supports approximately 140,188 jobs, and collects $15.2 billion for the local economy annually [30]. As of 2016, there were 35,000 hotel rooms in San Antonio. Of those 13,500 of them were in the downtown area [31]. In the third quarter of 2017, there were more than 2,000 active listings, making San Antonio the fourth highest-grossing city in Texas for Airbnb rentals [32].

Hotels give back to their communities by paying an average of $186 billion in local, state, and federal annual taxes combined in the U.S [33]. In San Antonio, the Hotel Occupancy Tax (HOT) revenue goes to support convention activities, public improvements, tourism, art and cultural programming for the city [34]. Up until 1 May 2017, a HOT was not collected from Airbnb's in Texas. In 2016, this resulted in an estimated $8 million HOT loss from Airbnb's in San Antonio alone [32]. Now that cities are starting to regulate STRs, there has been a decrease in Airbnb's.

4.1 Short-term rental regulations

In order to sustain tourism and prevent displacement, on 1 November 2018, San Antonio's City Council passed an ordinance establishing regulations over STRs, which included "the protection of the health and safety of occupant(s) of short-term rental properties, and to protect the integrity of the neighborhoods in which short-term rental properties operate" [35]. The city defines a STR as a "residential dwelling unit, apartment, condominium or accessory dwelling where sleeping areas are rented to overnight guests for a period of less than 30 consecutive days (but not less than 12 hours)" [36]. Under the ordinance STRs must apply for a permit with the COSA Development Services Department (DSD) in order to legally operate within San Antonio city limits. There are two types of permits:

1. Type 1 – The primary residence of the operator or owner; allowed anywhere if the operator or owner resides on the property.
2. Type 2 – The operator or owner does not occupy the residence; must follow density rules.

Density limitations are only for Type 2 STRS which can only be 12.5% of block-face in a residential area, by right, and in a multi-family housing unit can also only have no more than 12.5% of the total number of units be STRs, by right. Only the Board of Adjustments may allow the 12.5% density to be exceeded [37].

4.2 Short-term rentals in the south

Even though STRs are dense in the heart of downtown, this study focuses on an area south of downtown along the UNESCO San Antonio Missions World Heritage Buffer Zone (WHBZ). The zone consists of many neighborhood associations such as the Downtown Residents Association, King William, Lavaca, Lone Star, Roosevelt Park, Mission San Jose, Hot Wells Mission Reach, and Villa Coronado. Although each association and neighborhood are unique, all were created to preserve, protect, and sustain each community character. The Downtown Residents Association was created in 1982 and strives for a live work and play vibe [38]. King William's Association was created in 1967 and is mainly a neighborhood of Victorian homes built for prosperous German families after the Civil War [39]. Lavaca has a diverse culture that is highly active within the community. It is the oldest neighborhood in San Antonio [40]. The Lone Star Neighborhood Association is one the newest organizations in the city, founded in 2013, and is known for the artistic culture within their neighborhood [41]. Roosevelt Park Neighborhood Association was established in 1978 and main mission is to enhance the quality of life and create a safe environment within their neighborhood [42]. Mission San Jose Neighborhood Association was established in 1995 and consists of residents that are descendants of the laborers of Mission San Jose.

Aside from the Downtown Residents Association, King William, and Lavaca, the other neighborhoods in the WHBZ are predominately of a minority and lower income demographic, a reason why these neighborhoods are vulnerable to an increase in STRs. These "edgy" neighborhoods are trending because of the realness and authentic experience they provide and their strong intangible heritage.

In order to visualize the growth of STRs, current data were collected from City of San Antonio's (COSA) Development Services, and merged with shapefiles downloaded from the COSA GIS website. The study area was broken down into the closest geographical spatial data to the WHBZ, zip codes: 78204, 78210, 78214, and 78223. The WHBZ and these zip codes can be seen in Fig. 3, along with the number of short-term rentals within these zip codes. The ordinance has only been in full effect since February 2019. Only 519 STRs were grandfathered in, as of 11 April 2019. The city believes that there are about 1,500–2,000 properties that are not operating legally [43]. From the data provided by the City of San Antonio's Development Services Department, it was found that there is a total of 375 STRs in San Antonio, that the city is aware of, as of 24 February 2020. This includes grandfathered, non-compliant and registered STR properties.

Zip code 78210 consists of 63% of known STRs in San Antonio (236 properties). Out of these 236 properties, 146 are registered with the city and 76 properties have been grandfathered into the system. This means that approximately 6% of STRs in the 78210 zip code are non-compliant. Zip code 78204 has the second highest number of STRs with 112 properties, with 87 properties registered with the city, 13 grandfathered and 12 non-compliant.

Zip codes 78210 and 78204 are located nearest San Antonio's downtown core. The largest clusters reside furthest North of the study area which then trickle off alongside of the highways and South near the WHBZ boundary. Zip code 78223 is the zip code that stretches

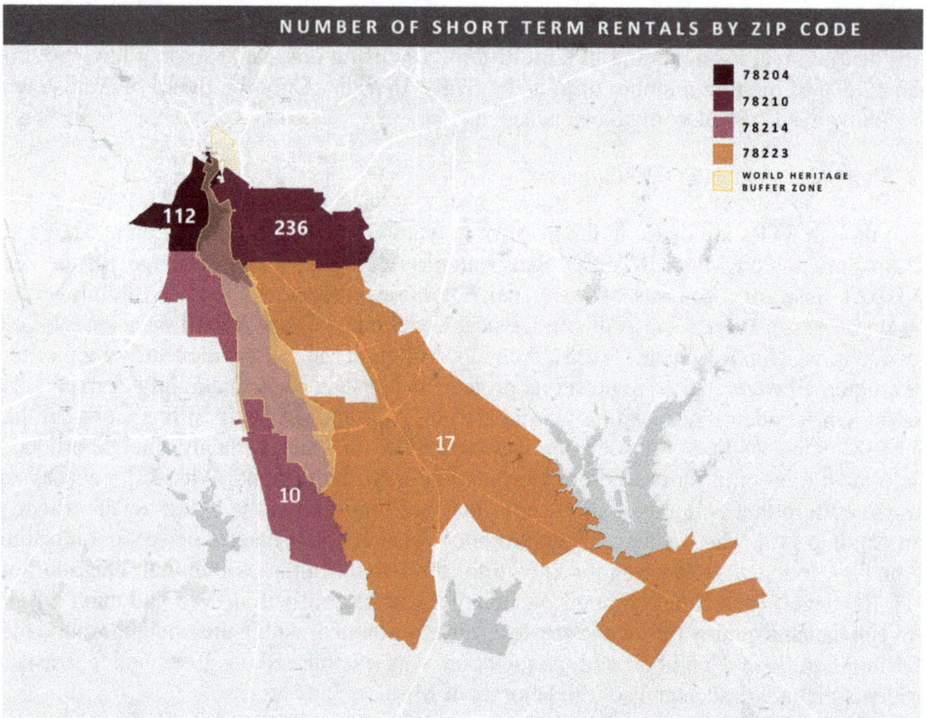

Figure 3: Short-term rentals by zip code as of 24 February 2020. Data sourced from City of San Antonio's Development Services Department. Chart created 30 April 2020 by Selina Angel.

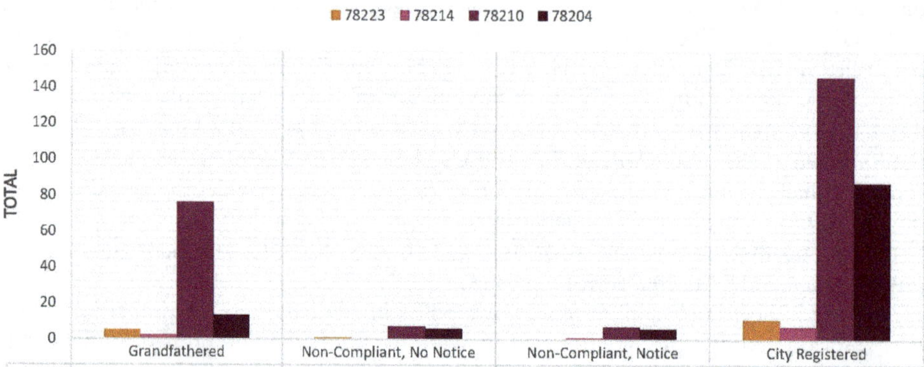

	Grandfathered	Non-Compliant, No Notice	Non-Compliant, Notice	City Registered
78223	5	1	0	11
78214	2	0	1	7
78210	76	7	7	146
78204	13	6	6	87

Figure 4: Short-term rental status as of 24 February 2020. Data sourced from City of San Antonio's Development Services Department. Chart created 30 April 2020 by Selina Angel.

furthest South but does not have the least amount of STRs. It is the 78214 zip code that has the least amount, at a total of 10 properties (seven registered with the city, two grandfathered and one non-compliant), however the 78214 STRs still follow along the WHBZ boundary while the 78223 STRs are seen furthest away from the WHBZ boundary (Figs 3 and 4).

5 CONCLUSION

There is an increase in development amongst the neighborhoods that encompass the San Antonio Missions leading to an increase of STRs in these well-established but quite vulnerable neighborhoods. In this case, STRs may not have had a significant impact on the social fabric of a community yet, however we know the housing market is drastically changing. For it is the unique, distinct character of place that brings STRs into existence, but it is also that same element of heritage and irreplaceable identity that must be protected. Fortunately, San Antonio has put in place regulations and is implementing STRs policies, which may help communities, stay intact, and residents not gentrified not only within in the World Heritage Buffer Zone but also the surrounding areas. This way, "shared economy" with STRs can be sustained while controlling the growth, and still preserving the history and identity of neighborhoods. Since this research was done at a time, where regulations were fresh, as time passes more information on STRs will be readily available. Additionally, future studies will have a greater amount of information and can determine if the regulations have caused a plateau in STRs or has caused an overall disappearance.

As previously mentioned, the social impacts of STRs take away the sense of a neighborhood. Communities become commodities and everyday life becomes difficult for residents because the neighborhood becomes a theme park for tourists or on the other side of the spectrum ghost towns. These negative social impacts may have been avoided with the help of new STR regulations, but further demographic and housing market studies should be also be completed in order to fully understand if there are any significant correlations between these three variables. All of these data may also drastically change after COVIT19 pandemic in 2020, and the change in both tourist demand and residents' response should be analyzed in the future.

REFERENCES

[1] City of San Antonio (COSA), *World Heritage Zone Neighborhood, Community And Sector Plan Update,* 30 pp., 2016. www.sanantonio.gov/Portals/0/Files/WorldHeritage/WorkPlanUpdates/LandUse/ENGLISH%202016%20WHBZ%20City%20Approved%20Land%20Use.pdf?ver=2018-02-05-101757-203. Accessed on: 30 Jan. 2020.

[2] National Park Service, *2018 National Park Visitor Spending Effects,* National Resource Stewardship and Science: Fort Collins, CO, 57 pp., 2019.

[3] Hosking, N., Why we all need to travel with purpose, Global Heritage Fund, 2019. https://globalheritagefund.org/2019/09/27/why-we-all-need-to-travel-with-purpose/.

[4] U.S. Travel Association, U.S. travel answer sheet, 2019. www.ustravel.org/answersheet. Accessed on: 3 Feb. 2020.

[5] Payne, J., Insider exclusive: Heritage tourism – Facts and figures, American Bus Association, 2018. www.buses.org/news/article/insider-exclusive-heritage-toursim-facts-figures. Accessed on: 4 Feb. 2020.

[6] World Tourism Organization, *Tourism and Culture Synergies*, The (UNWTO), 2018. www.e-unwto.org/doi/pdf/10.18111/9789284418978. Accessed on: 4 Feb. 2020.

[7] Barcelona Field Studies Centre S. L., Tourism multiplier effect, 2020. https://geographyfieldwork.com/TouristMultiplier.htm. Accessed on: 9 Feb. 2020.

[8] Doganer, S. & Dupont, W., Accelerating cultural heritage tourism in San Antonio: A community-based tourism development proposal for the Missions Historic District. *International Journal of Sustainable Development Planning*, **10**(1), pp. 1–19, 2015.
[9] Jun, S.H., Nicholls, S. & Vogt, C., Heeding the call for heritage tourism: more visitors want an "experience" in their vacations: something a historical park can provide, 2004. http://citeseerx.ist.psu.edu/viewdoc/download?doi=10.1.1.602.6461&rep=rep1&type=pdf.
[10] Martinelli, K., Everything you need to know about the vacation rental industry, Part 1, 2020. www.futurestay.com/read/vacation-rental-history/vacation-rental-industry/.
[11] Akande, I., The history of vacation rentals: Infographic, 2015. https://rentalsunited.com/blog/history-of-vacation-rentals-infographic/. Accessed on: 11 Feb. 2020.
[12] VRM Intel Staff, Measuring performance for vacation rental management companies, 2013. www.vrmintel.com/part-one-a-look-back-at-how-the-marketing-and-competitive-landscape-has-changed-for-vacation-rental-managers/. Accessed on: 16 Feb. 2020.
[13] KeyCafe, The history of short-term rentals, 2020. https://blog.keycafe.com/the-history-of-short-term-rentals/. Accessed on: 18 Feb. 2020.
[14] Jasper, The Airbnb founder story: From selling cereals to a $25B company, 2019. https://getpaidforyourpad.com/blog/the-airbnb-founder-story/. Accessed on: 18 Feb. 2020.
[15] iProperty Management, Airbnb statistics, 2020. https://ipropertymanagement.com/research/airbnb-statistics. Accessed on: 22 Feb. 2020.
[16] Condor Ferries Ltd., Top hotel statistics, 2000. www.condorferries.co.uk/hotel-industry-statistics. Accessed on: 24 Feb. 2020.
[17] Gessner, K., Ahead of IPO, Airbnb's consumer sales surpass most hotel brands, Second measure, 2019. https://secondmeasure.com/datapoints/airbnb-sales-surpass-most-hotel-brands/.
[18] Guttentag, D., Airbnb: Disruptive innovation and the rise of an informal tourism accommodation sector. *Journal of Current Issues*, **18**(12), pp. 119–1217, 2013.
[19] @Silver86, Airbnb's brand evolution, 2016. https://all-about-airbnb.com/post/143221488726/airbnb-brand-evolution-live-there. Accessed on: 25 Feb. 2020.
[20] Airbnb, New data: The Airbnb advantage, 2019. https://news.airbnb.com/new-data-the-airbnb-advantage/. Accessed on: 4 Apr. 2020.
[21] Gisolf, M., The background of sustainable tourism, 2010. www.tourismtheories.org/?p=958. Accessed on: 6 Apr. 2020.
[22] Wisniewska, A., Are Airbnb investors destroying Europe's cultural capitals? *Financial Times*, 2019. www.ft.com/content/2fe06a7c-cb2a-11e9-af46-b09e8bfe60c0.
[23] Jane Place Neighborhood Sustainability Initiative, Short-term rentals, long-term impacts: The corrosion of housing access and affordability in New Orleans. https://storage.googleapis.com/wzukusers/user27881231/documents/5b06c0e681950W9RSePR/STR%20Long-Term%20Impacts%20JPNSI_4-6-18.pdf.
[24] Geier, B., Where investors are taking advantage of Airbnb 2019, CEPF, 2019. https://smartasset.com/financial-advisor/where-investors-are-taking-advantage-of-short-term-rental-market-2019. Accessed on: 8 Apr. 2020.
[25] Judd, D.R., Visitors and the spatial ecology of the city. *Cities and Visitors: Regulating People, Markets and City Space,* eds L.M. Hoffman, S.S. Fainstein & D.R. Judd, Blackwell Publishing, pp. 23–38, 2003. https://onlinelibrary.wiley.com/doi/10.1002/9780470773673.ch.

[26] Cocola-Grant, A. & Gago, A., Airbnb, buy-to-let investment and tourism-driven displacement: A case study in Lisbon. *Environment and Planning A: Economy and Space,* pp. 1–18, 2019.
[27] CBRE, Hosts with multiple units: a key driver of airbnb growth. a comprehensive national review including a spotlight on 13 U.S. markets, 2017. www.ahla.com/sites/default/files/CBRE_AirbnbStudy_2017_0.pdf.
[28] Dolven, T., Hanks, D. & Gurney, K., Why Airbnb is making it harder for Miami locals to find a place to rent, 2019. www.miamiherald.com/news/business/tourism-cruises/article231476068.html. Accessed on: 9 Apr. 2020.
[29] Egan, J., Alamo City crowned World's top tourist destination for 2018 once again, Culturemap San Antonio, 2017. https://sanantonio.culturemap.com/news/city-life/12-11-17-travel-leisure-magazine-touts-san-antonio-worlds-top-travel-destinations-for-2018/. Accessed on: 9 Apr. 2020.
[30] Visit San Antonio, Celebrate national travel and tourism week: May 5–11, 2019, 2019. www.visitsanantonio.com/nttw/. Accessed on: 13 Apr. 2020.
[31] Rivard, R., Getting real with the San Antonio visitor economy, 2016. https://therivardreport.com/getting-real-san-antonio-visitor-economy/. Accessed on: 13 Apr. 2020.
[32] Dimmick, I., Short-term rentals debate to continue at City Hall, State Lege. The Rivard Report, 2017. https://therivardreport.com/short-term-rentals-debate-to-continue-at-city-hall-state-lege/. Accessed on: 27 Mar. 2020.
[33] Oxford Economics, Economic impact of the US hotel industry, 2019. www.ahla.com/sites/default/files/oxford2019.pdf. Accessed on: 23 Mar. 2020.
[34] City of San Antonio, Hotel occupancy taxes, 2000–2020. www.sanantonio.gov/Finance/about/faq. Accessed on: 20 Mar. 2020.
[35] City of San Antonio, November 1. Article XXII. Short term rentals division I: general, 2018. https://docsonline.sanantonio.gov/DSDUploads/STROrdinanceAsPassed.pdf. Accessed on: 14 Apr. 2020.
[36] City of San Antonio, Short-term rentals ordinance fact sheet, 2019. https://docsonline.sanantonio.gov/DSDUploads/STRFactSheet.pdf. Accessed on: 16 Apr. 2020.
[37] City of San Antonio, Indoor notification template, 2019. https://docsonline.sanantonio.gov/DSDUploads/STRIndoorNotificationTemplate.pdf. Accessed on: 19 Apr. 2020.
[38] Meyer, C., DRA annual activities March 2008–March 2009. www.neighborhoodlink.com/Downtown/info. Accessed on: 27 Apr. 2020.
[39] King William Association, Who we are, 2020. www.kwfair.org/p/about/147. Accessed on: 27 Apr. 2020.
[40] Lavaca Neighborhood Association, Welcome to Lavaca! 2011. http://www.lavaca.net/. Accessed on: 28 Apr. 2020.
[41] Southtown the Arts District, About District Mission, 2018. https://southtownsatx.com/about/. Accessed on: 28 Apr. 2020.
[42] Roosevelt Park Neighborhood Association, Inspiring a neighborhood, 2020. www.rooseveltparkna.org/about/. Accessed on: 28 Apr. 2020.
[43] Alley, L.T. & Rivers, R.D., Short-term rental study 2019, ed. City of New Orleans, City Planning Commission, 2019. www.nola.gov/nola/media/City-Planning/Final-2019-STR-Study-with-Maps.pdf. Accessed on: 23 Apr. 2020.

TOURIST BEHAVIOUR AND TICK-BORNE DISEASE RISK

ALYSSA SOUCY & SANDRA DE URIOSTE-STONE
School of Forest Resources, University of Maine, USA

ABSTRACT
Lyme disease is the leading tick-borne disease in the USA, with incidences increasing over the past few decades due in part to climate change. Visitors to natural areas involved in nature-based activities are at a heightened risk of Lyme disease exposure. We surveyed 430 visitors in Acadia National Park to understand the personal protective behaviours and associated barriers to adoption, perceived tick-borne disease risk, and travel behaviour. A two-step cluster analysis revealed two visitor segments: adventurists and sightseers. Adventurists were less likely to perform protective behaviours against tick-borne disease, but performing a tick check was the most commonly reported preventative behaviour for both groups (62%). The most commonly cited barriers to performing a tick check were lower perceived risk of tick bites and Lyme disease for both groups. Both groups also reported that the number one barrier to wearing protective clothing was the hot summer weather. This research has implications for tourism managers to design effective communication materials to reduce the risk of tick-borne disease for different visitor segments.
Keywords: nature-based tourism, cluster analysis, Lyme disease, Acadia National Park, ticks, health risk, survey research, Maine.

1 INTRODUCTION

Lyme disease (LD), transmitted through the bite of an infected blacklegged tick, is the most commonly diagnosed tick-borne disease (TBD) in the United States, with over 42,000 reported cases in 2017 [1]. Lyme disease poses a significant threat to those travelling to rural areas partaking in outdoor activities [2]. Risks may be heightened for tourists unaware of LD who lack the necessary knowledge and resources to properly protect themselves [3]. On the other hand, impacts from localized climate change, such as increases in LD risk, can impact the quality of the visitor experience [4] and their decisions regarding future travel plans [5]. For that reason, exposure to LD has major implications for both visitors and tourism managers. The purpose of this study is to analyse the role of activity type and associated factors that may influence personal protective behaviours against TBD in Acadia National Park, Maine, United States of America.

Tick-borne disease risk management primarily focuses on protective measures to reduce TBD risk. Personal protective measures include using insect repellents containing DEET, avoiding contact with ticks, performing tick checks after recreating outdoors, and showering within 2 hours of coming indoors [6]. Despite the wide range of protective measures against TBD and high knowledge of LD symptoms, uptake has been poor [7], [8]. Large differences exist in the type of protective measure used among the public, with performing a tick check as the most commonly reported behaviour [9], [10] and using insect repellent as the least commonly reported [11]. Similarly, there have been disparities in the effectiveness of education to increase protective behaviours [3], [12]. Hence, understanding the determinants of protective measure adoption is important to addressing this health risk.

Potential factors that may influence the uptake of protective behaviours against TBD have been extensively studied. Perceived efficacy of a given protective measure has been shown to be strongly associated with performing that measure [10]. Knowledge and concern are also significant predictors of protective behaviour [13], [14]. There do, however, exist differences in the potential drivers for the use of different protective measures. Knowledge about LD

www.witpress.com, ISSN 1743-3541 (on-line)
doi:10.2495/ST200071

[15] and exposure to ticks [16] have been significantly associated with tick checks but not with other measures such as tick repellent [15].

In tourism studies the use of segmentation analysis can help compare groups of visitors with different characteristics to identify variations that exist between these groups [17] in terms of travel or health behaviours. Acadia National Park brings visitors from all over the globe who engage in a variety of recreational activities; therefore, determining if differences exist in how diverse visitor groups perceive and respond to the risk of TBD can help guide risk communication strategies that seek to increase adoption of personal protective behaviours. The goal of this study is to (a) investigate the differences in socio-demographics, perceived risk, and personal protective behaviours by visitor group; and (b) identify barriers to adopting protective behaviours by visitor group.

2 METHODS

2.1 Study area

Acadia National Park (ANP) is located in Maine, which is in the north-eastern part of the United States. Over the past several decades there have been increases in incidences of LD due in part to climate change [18]. In 2017, Maine had 1,852 Lyme diagnoses, making it the state with the highest amount of confirmed Lyme cases per 100,000 residents [19]. Tourism is one of the largest industries in the state of Maine. In particular, ANP offers a diverse range of recreation activities that attract over 3 million visitors every year, including a scenic road that loops through the entire park, guided programs and museums, more than 150 miles of trails for all skill-levels, campgrounds, and opportunities for rock climbing, kayaking, bird watching [20]. ANP was selected as our study site for the following reasons: (1) the park is the 8th most visited National Park in the United States, with outdoor recreation as primary attraction, hence putting millions of tourists per year at risk of exposure TBD; (2) TBD is expected to increase as a result of climate change and increases visitation in the study area; (3) visitation coincides with peak periods of black-legged tick activity; and (4) a 2016 study found that ANP visitors identified heightened risk of exposure to vector-borne disease as an important concern regarding future visitation [5] that may have economic implications for nearby gateway communities.

2.2 Survey design and sampling procedure

The study used a mixed-mode survey, with an on-site 5-minute intercept component followed by a longer online self-administered survey. The intercept survey included socio-demographic and travel behaviour questions as well as questions regarding perceived TBD risk, and was aimed at increasing the likelihood that visitors would complete the longer self-administered survey. The self-administered instrument consisted of questions on travel behaviour, and protective behaviours to reduce TBD risk. Data were collected between May–October 2018 in ANP.

Participants were selected using a two-stage cluster probability sample [21]. The sampling dates were first randomly selected and the visitor groups were then selected on-site using a systematic probability sampling technique [22]. Survey administrators approached random walking groups by using a pre-selected interval (i.e. every 3rd group), and interviewed the person with the most recent birthday, if willing. Following the short on-site survey, participants were handed a postcard with a link to a self-administered online survey to complete once they returned home. Dillman's Tailored Design Method was used to maximize response rate by asking visitors for an email or mailing address in order to send up to two reminders regarding the online survey [23]. Of the 1,252 on-site surveys collected, 430 respondents completed the self-administered survey (34% response rate).

2.3 Measures

2.3.1 Activities

Respondents were asked to select which activities they participated in during their visit to ANP out of a list of 33 options. Respondents were then asked to select their primary activity from that same list of options. Participants were also asked if they would make changes in their future travel plans based on TBD concerns (change destination, engage in different recreational activities, change the timing of their travel).

2.3.2 Perceived risk

Respondents were asked to rate their responses to the question of whether or not the risks that ticks pose to humans are (1) *avoidable* and (2) *controllable* using 7-point Likert scale where 1 = strongly agree and 7 = strongly disagree. The two questions were combined using a mean score, where answers to both questions were required, so that higher scores indicate higher perceived risk. These two questions were highly correlated with a Pearson's correlation of 0.60.

2.3.3 Protective behaviour

Participants were asked to select whether or not they engaged in 8 personal protective behaviours *never, sometimes,* or *always.* Protective behaviours included *staying on trails, performing tick checks, tucking pants into socks, avoiding activities that expose one to ticks, avoiding recreating outdoors, using insect repellent, wearing light-coloured clothing,* and *wearing protective clothing.* Responses were coded so that 0 = *never*, 1 = *sometimes*, and 2 = *always.* A sum score was calculated so that higher scores reflect higher overall protective behaviour ($\alpha = 0.76$). Scores are out of 16.

2.3.4 Barriers to protective behaviours

If a participant did not select always for the following protective behaviours: performing tick checks or wearing protective clothing, they were asked to select their top three barriers to that behaviour out of a list of 5 options. Both behaviours listed: there is a low chance of getting Lyme disease, there is a low chance of getting bit by a tick, and that behaviour is overdone as potential barriers. Other barriers for performing tick checks included not knowing how to remove a tick and not knowing how to recognize a tick. Responses were coded so that 1 indicates a top barrier and 0 indicates that that is not a top barrier.

2.3.5 Socio-demographics

Lastly, a range of socio-demographics was collected including age, gender (0 = female; 1 = male), education collapsed into two categories where 0 = higher education (college degree); 1 = lower education (no college degree), their home state/country (1 = states with LD; 2 = all other), and whether or not they are a 1st time visitor to ANP (0 = 1st time visitor; 1 = repeat visitor).

2.4 Data analysis

There were a total of 430 responses to the self-administered survey. Non-response bias was measured using Pearson's chi-square test of independence (χ^2) to compare those who responded to the on-site survey ($n = 1252$) with those who completed the self-administered survey ($n = 429$). There was no statistical differences in first time visitation ($\chi^2 = 1.45$, 1 df, $p = 0.23$), knowledge of what a tick is ($\chi^2 = 2.85$, 1 df, $p = 0.09$), and gender ($\chi^2 = 2.44$, 1 df, $p = 0.12$); however, there was a statistical difference in experience with TBD ($\chi^2 = 5.31$, 1 df, $p = 0.02$) with those taking the on-site survey having relatively fewer experience with TBD compared to those taking the self-administered survey. It is possible that those who have personally experienced TBD were more interested in the study and therefore were more inclined to take the self-administered survey, which would account for this difference.

A multivariate two-step cluster analysis was run to segment ANP visitors. Cluster analysis can be useful in the context of tourism and parks management as means to understand visitor group needs [24]. Clustering was based off of the number of nature-based activities (either somewhat nature-based or nature-based) and whether the primary activity was nature based (Table 1) as modified by the procedure used in Wilkins et al. [25].

Independent samples t-tests were used to test for differences for the continuous variables of perceived risk, and protective behaviour. Levene's statistic was used to test the assumption of equal variances of groups. If homogeneity of variance was violated an adjustment was made using the Welch-Satterthwaite method [26]. Cohen's d was used to assess effect size for independent samples and Welch's t-test results. Chi-square tests were run to examine the differences in activity groups for each protective behaviour. Cramer's V was used for effect size and adjusted standardized residuals (ASR) were used as a post-hoc test, with those two standard deviations or more away from the expected mean reported. Barriers to protective behaviours against TBD were investigated using descriptive statistics. The proportion of respondents who selected each barrier was calculated for each group using the number of respondents who did not *always* perform that protective behaviour. All data analyses were done in SPSS 25.0.

Table 1: Categorization of types of activities listed used for two-step cluster analysis.

Categories	Activities
Not nature-based (0)	Arts or cultural activity, Concert or festival, Dining at Jordan Pond House
Sightseers (1)	Sightseeing or driving for pleasure
Somewhat nature-based (2)	Bird Watching, Going to the beach, Picking berries, Picnicking, Taking horse or carriage ride, Trail running, Walking my dog, Wildlife viewing
Nature-based (3)	Backpacking, Biking, Camping, Hiking, Kayaking, Non-technical mountain climbing, Walking on carriage roads, Swimming

3 RESULTS

3.1 Demographic profile, trip characteristics, and potential travel behaviour change

As shown in Table 2, the majority of respondents were non-residents of Maine (93%) with 54% from areas with LD and 39% from areas without LD. Most respondents (61%) were first time visitors, travelling with family (78%). The 60–69 year age group had the most respondents (28%) followed by the 50–59 age group (22%) and then the 40–49 age group (20%). Finally, there was a majority female sample (60%). In addition, almost half of the respondents selected hiking on trails (49%) as their primary activity with sightseeing as the second most selected response (27%), and biking as the third (7%). Over a quarter of respondents participated in the following activities at ANP: hiking on trails (76%), sightseeing/driving for pleasure (75%), going to the beach (40%), viewing wildlife (39%), shopping in the park (36%), and dining at Jordan Pond House (28%). Further, when asked about changes in travel behaviour resulting from concerns with TBD, 35% of participants expressed that they would change their travel destination, 26% would change their outdoor recreation activities, while 17% would consider traveling during another time of year.

Table 2: Socio-demographics, trip characteristics and potential travel behaviour change.

Variable	**Subcategory**	**No. of responses**	**Percent**
Residency	Maine	25	7%
	Areas with Lyme Disease	187	54%
	All other areas	137	39%
First time visitation to ANP	First time visitors	253	61%
	Repeat visitors	164	39%
Type of personal group	Alone	15	4%
	Family	296	78%
	Friends	44	12%
	Family and Friends	25	7%
Age (years)	< 30	27	7%
	30–39	55	15%
	40–49	72	20%
	50–59	80	22%
	60–69	101	28%
	> 69	32	8%
Gender	Male	154	40%
	Female	229	60%
Travel behaviour change from TBD concern	Change destination	135	35%
	Change in outdoor activities	98	26%
	Change time of year to travel	66	17%

3.2 Recreation activity groups and their differences

The clustering analysis resulted in two visitor segments: adventurists and sightseers (Table 3). Adventurists were those who participated in more nature-based activities ($M = 4.5$) and whose primary activity was nature-based, measured on a scale from 0–3 ($M = 3.0$). Sightseers were those who participated in fewer nature-based activities ($M = 2.5$) and whose primary activity was less-nature based ($M = 1.2$). Table 4 shows there is no significant difference in gender (χ^2 (1, $N = 374$) = 1.21, $p = 0.27$) or first time visitation (χ^2 (1, $N = 303$) = 2.52, $p = 0.11$). Education and region were significantly different between activity groups, with adventurists tending to be more highly educated (χ^2 (1, $N = 374$) = 1.21, $p = 0.27$) and from areas with the presence of LD (χ^2 (1, $N = 350$) = 8.58, $p = 0.00$).

As shown in Table 5, t-tests elicited no significant differences between activity groups in perceived risk ($t(1) = 0.15$, $p = 0.88$). Overall, sightseers ($M = 7.97$) reported performing protective behaviours against TBD more often than adventurists ($M = 7.27$) on a scale from 0 to 16 ($t(1) = -2.02$, $p = 0.03$); the overall protective behaviour comprises 8 measures.

Table 3: Descriptions and characteristics of the two activity visitor groups (clusters).

Visitor group	Adventurists	Sightseers
Description	Tourists who tended to participate in more nature-based activities and whose primary activity was nature-based, like hiking, camping, etc.	Tourists who tended to participate in fewer nature-based activities and whose primary activity was less nature-based, such as sightseeing
Size	260 (65%)	143 (35%)
Inputs	Mean number of nature-based activities: 4.49 Primary activity nature-based: 3.0	Mean number of nature-based activities: 2.53 Primary activity nature-based: 1.15

Table 4: Comparisons of socio-demographics and trip characteristics broken down by activity group (reported as percentages).

Variable	Adventurists (%)	Sightseers (%)	Chi-square (sig)	Cramer's V (Φ_C)
Gender	*N* = 236	*N* = 138		
Male	38	43	1.21 (0.27)	0.06
Female	62	57		
Home state/country	*N* = 219	*N* = 131		
Areas with LD	66	50	8.58 (0.00)	0.16
All other areas	44	50		
Education	*N* = 238	*N* = 138		
Lower education	8	20	10.91 (0.00)	0.17
Higher education	92	80		
Visitation	*N* = 260	*N* = 143		
First time visitor	57	65	2.52 (0.11)	0.08
Repeat visitor	43	35		

Table 5: Comparisons of perceived risk and protective behaviour broken down by activity group (reported as mean values).

Variable	Adventurists (N)	Sightseers (N)	Levene Stat (sig)	t-test (sig)	Cohen's D
Perceived risk	2.65 (241)	2.63 (138)	1.58 (0.21)	0.15 (0.88)	0.01
Protective behaviour	7.27 (244)	7.97 (137)	7.79 (0.01)	-2.02 (0.03)	0.24

When examining individual protective behaviours, there is no significant difference in performing a tick check (χ^2 (2, N = 379) = 5.36, p = 0.07) or using insect repellent (χ^2 (2, N = 370) = 1.38, p = 0.50) between adventurists and sightseers as shown in Table 6. There are, however, a relatively larger percentage of sightseers (8%) who *never* perform a tick check compared to adventurists (3%). Despite this, performing a tick check is the most commonly performed protective behaviour for both groups (62% always perform a tick check). The significant difference in overall protective behaviour can be attributed to

Table 6: Comparisons of types of protective behaviours against TBD broken down by activity group (reported as percentages).

Variable	Overall (%)	Adventurists (%)	Sightseers (%)	Chi-square (sig)	Cramer's V (Φ_C)
Performing a tick check	*N* = 387	*N* = 244	*N* = 135	5.36 (0.07)	0.20
Never	5	3*	8*		
Sometimes	33	34	32		
Always	62	63	60		
Wearing protective clothing	*N* = 384	N = 242	N = 134	14.59 (0.00)	0.20
Never	9	10	8		
Sometimes	66	72*	56*		
Always	25	18*	36*		
Using insect repellent	*N* = 388	*N* = 244	*N* = 136	1.38 (0.50)	0.06
Never	9	9	7		
Sometimes	54	56	52		
Always	37	35	40		
Avoiding activities that will expose to ticks	*N* = 388	*N* = 244	*N* = 135	9.84 (0.01)	0.16
Never	41	41	42		
Sometimes	54	57	50		
Always	5	2*	8*		
Avoiding recreating outdoors	*N* = 385	*N* = 243	*N* = 133	6.37 (0.04)	0.13
Never	41	83*	74*		
Sometimes	54	16	23		
Always	5	1	3		
Staying on paved trails/gravel	*N* = 386	*N* = 244	*N* = 134	6.07 (0.05)	0.13
Never	14	15	12		
Sometimes	67	70	63		
Always	19	15*	25*		
Tucking pants into socks	*N* = 386	*N* = 244	*N* = 15	11.15 (0.01)	0.17
Never	41	46*	35*		
Sometimes	44	44	43		
Always	15	10*	22*		
Wearing light coloured clothing	N = 385	N = 242	N = 135	6.19 (0.05)	0.13
Never	19	19	20		
Sometimes	66	70*	60*		
Always	15	11*	20*		

Note. Degrees of freedom = 2 for all chi-square tests.
*Indicates adjusted standardized residual (ASR) > 1.96.

sightseers more commonly performing other protective behaviours, such as wearing protective clothing (χ^2 (2, $N = 376$) = 14.59, $p = 0.00$).

3.3 How do the barriers to protective behaviours differ between activity groups?

Compared with only 64% of sightseers, 82% of adventurists do not *always* wear protective clothing. The top overwhelming barrier for both activity groups is that it is too warm in the summer, with 93% of all respondents selecting it as one of their top 3 barriers (Fig. 1). Both adventurists and sightseers prioritized similarly other barriers to using protective clothing as a protective measure.

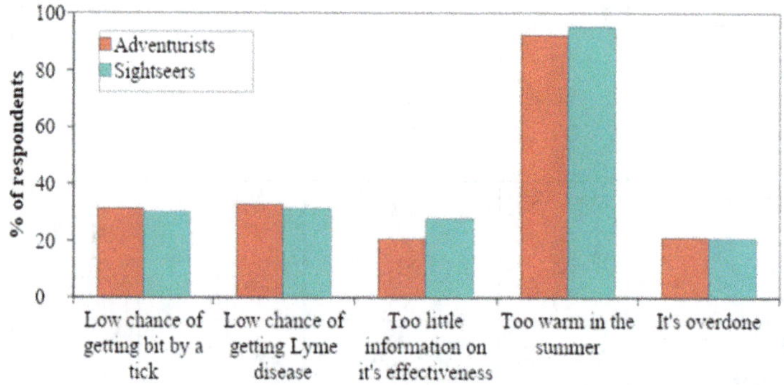

Figure 1: Bar chart of proportion of respondents who selected each barrier to wearing protective clothing for adventurists (pink) and sightseers (blue).

Approximately 40% of both activity groups do not *always* perform tick checks (Fig. 2). Adventurists selected *there's a low chance of getting Lyme* as their number one barrier to performing tick checks and *there's a low chance of getting bit by a tick* as their second barrier with 42% and 41% respectively. Sightseers were similar in that they selected both as their top barriers; however, 48% selected *there's a low chance of getting bit by a tick* and only 32% selected *there's a low chance of getting Lyme* as their top barriers to performing tick checks. *It's overdone* was the next highest barrier for adventurists with 19% of respondents

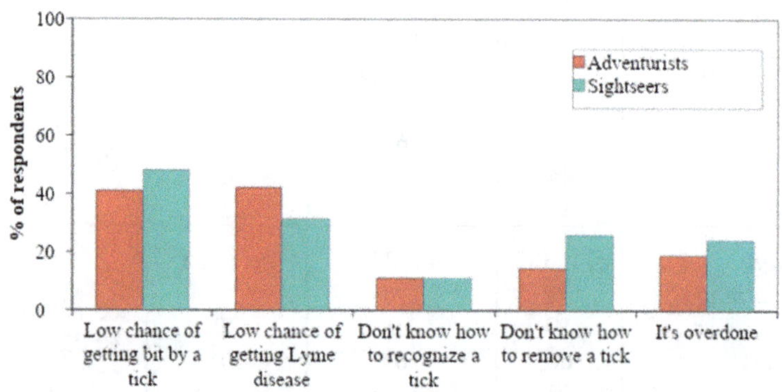

Figure 2: Bar chart of proportion of respondents who selected each barrier to performing a tick check for adventurists (pink) and sightseers (blue).

selecting it, while *I don't know how to remove a tick* was the next highest barrier for sightseers with 26% choosing it as a top 3 barrier. *I don't know how to recognize a tick* came in last at 11% for all respondents.

4 DISCUSSION AND CONCLUSION

Two-step cluster analysis was an effective way to group visitors to ANP based on activity type, with those engaging in more nature-based activities such as hiking and camping (adventurists) falling into one group, while those that primarily sightsee and drive the Park Loop Road (sightseers) fall into another group. Adventurists were more likely to be highly educated and from areas with LD, while sightseers were slightly less educated and from more areas without LD. Despite the differences in socio-demographics, there were no significant differences in perceived risk of ticks between activity groups. Overall, ANP visitors surveyed tend to have low perceptions of risk, with close to 10% of respondents perceiving that risks that ticks pose are avoidable and controllable; findings are in contrast to results from a prior study in Sweden [11] that found that 43% of respondents perceive tick bites as a serious health risk.

Further, differences were found in terms of preventive behaviours, with adventurists having lower overall protective behaviour than sightseers. This is due in part to the types of behaviours measured to reduce TBD risk, such as avoiding recreating outdoors and staying on gravel/trails. It is no surprise that those that engage in more nature-based activities while visiting ANP tend to do some of those behaviours less than those who sightsee regardless of whether or not the behaviour is truly done to avoid ticks. This may also be due in part to a psychological drive to take risks to experience a rush or thrill [27], or their belief that specific TBD protective behaviours might hinder their enjoyment of their preferred recreational activities [28]. Those that engage in nature-based activities at ANP may be less-inclined to wear protective clothing or stay on trails because that would be in opposition to their desire to freely explore the natural setting. It is, however, important to note that both adventurists and sightseers are regularly performing tick checks (62%), which is consistent with previous studies [8], [13]. Further, adventurists are wearing protective clothing less and more often not staying on trails – two protective measures specifically aimed at reducing TBD risk. From a management perspective, this is an important finding given that adventurists are inherently at higher risk of TBD due to the nature of their activities.

The top barrier to wearing protective clothing for both activity groups was that it is too warm in the summer; this finding is similar to those from previous studies [13], [29]. Other top barriers for wearing protective clothing and performing a tick check were a perceived low risk of getting bit by a tick and contracting Lyme disease; these compares findings from previous studies [13]. While it may be difficult to target warm weather as a barrier, increasing visitor's perceptions about the true risks of getting bit by a tick and contracting LD may increase preventative behaviours. More sightseers expressed their inability to remove a tick as a barrier to performing a tick check. While adventurists were relatively more knowledgeable about this, understanding the proper way to remove a tick may be an important way to increase tick checks in sightseers.

Although knowledge can inform risk perceptions and behaviours, research is needed to better understand the factors that determine perceptions of risk, and the drivers and barriers to visitor adoption of TBD protective behaviours while visiting natural settings, like national parks. Factors such as self-efficacy [8], [10], [30]; disgust about ticks [30]; perceived likelihood of being bitten by a tick [14], [30]; likelihood and severity of contracting a TBD [8], [13]; and barriers to adoption [13], may be important in predicting protective behaviours.

The results of this study demonstrate the existence of two main visitor groups to ANP: adventurists and sightseers. Sightseers perform more protective behaviours compared to adventurists. The differences in overall protective behaviour may be due to a variety of factors including the behaviours measured in the survey instrument or adventurists inherent inclination to take more risks; however, performing a tick check is relatively high for both groups. Targeting other protective behaviours such as using insect repellent or wearing protective clothing may require communication strategies that include information about how to properly remove ticks (aimed at sightseers) to increase self-efficacy [30], and the actual risk of contracting a TBD (aimed at adventurists). Risk communication strategies targeted to specific group characteristics will be important to increase risk perception and adoption of protective behaviours [31]. Further research can examine the drivers of protective behaviours in both groups and explore the role of barriers and self-efficacy in preventing certain types of protective behaviours.

ACKNOWLEDGEMENTS

This research was supported by the University of Maine Senator George J. Mitchell Center, National Oceanic and Atmospheric Administration under grant NA17OAR4310249, and National Science Foundation under grant 1828466. The authors thank study participants; staff from the National Park Service and Acadia National Park for their support; and UMaine undergraduate students – Asha DiMatteo-LePape, Nathaniel Burke, Hope Kohtala – and Lydia Horne for their dedication to data collection.

REFERENCES

[1] Centers for Disease Control and Prevention, Tickborne disease surveillance data summary. www.cdc.gov/ticks/data-summary/index.html. Accessed on: 14 May 2020.

[2] Donohoe, H., Pennington-Gray, L. & Omodior, O., Lyme disease: Current issues, implications, and recommendations for tourism management. *Tourism Management*, **46**, pp. 408–418, 2015.

[3] Daltroy, L.H., Phillips, C., Lew, R., Wright, E., Shadick, N.A. & Liang, M.H., A controlled trial of a novel primary prevention program for Lyme disease and other tick-borne illnesses. *Health Education and Behavior*, **34**, pp. 531–542, 2007.

[4] Brownlee, M.T.J., Hallo, J.C. & Krohn, B.D., Botanical garden visitors' perceptions of local climate impacts: Awareness, concern, and behavioral responses. *Managing Leisure*, **18**(2), pp. 97–117, 2013.

[5] De Urioste-Stone, S.M., Le, L., Scaccia, M.D. & Wilkins, E., Nature-based tourism and climate change risk: Visitors' perceptions in Mount Desert Island, Maine. *Journal of Outdoor Recreation and Tourism*, **13**, pp. 57–65, 2016.

[6] Centers for Disease Control and Prevention, Preventing ticks bites. www.cdc.gov/ticks/avoid/on_people.html. Accessed on: 15 May 2020.

[7] Valente, S.L., Wemple, D., Ramos, S., Cashman, S.B. & Savageau, J.A., Preventive behaviors and knowledge of tick-borne illnesses: Results of a survey from an endemic area. *Journal of Public Health Management and Practice*, **21**(3), 2015.

[8] Shadick, N.A., Daltroy, L.H., Phillips, C.B., Liang, U.S. & Liang, M.H., Determinants of tick-avoidance behaviors in an endemic area for Lyme disease. *American Journal of Preventive Medicine*, **13**(4), pp. 265–270, 1997.

[9] Butler, A.D., Sedghi, T., Petrini, J.R. & Ahmadi, R., Tick-borne disease preventive practices and perceptions in an endemic area. *Ticks and Tick-Borne Diseases*, **7**(2), 331–337, 2016. https://doi.org/10.1016/j.ttbdis.2015.12.003.

[10] Aenishaenslin, C. et al., Factors associated with preventive behaviors regarding Lyme disease in Canada and Switzerland: A comparative study. *BMC Public Health*, **15**(285), pp. 1–10, 2015.

[11] Slunge, D. & Boman, A., Learning to live with ticks? The role of exposure and risk perceptions in protective behaviour against tick-borne diseases. *PloS One*, **13**(6), 2018.

[12] Hayes, E.A. & Piesman, J., How can we prevent Lyme disease? *New England Journal of Medicine*, **348**, pp. 2424–2430, 2003.

[13] Beaujean, D., Bults, M., Steenbergen, J. & Voeten, H., Study on public perceptions and protective behaviors regarding Lyme disease among the general public in the Netherlands: Implications for prevention programs. *BMC Public Health*, **13**(1), pp. 225–225, 2013.

[14] Herrington, J.E., Risk perceptions regarding ticks and Lyme disease. *American Journal of Preventive Medicine*, **26**(2), pp. 135–140, 2004.

[15] Butler, A.D., Sedghi, T., Petrini, J.R. & Ahmadi, R., Tick-borne disease preventive practices and perceptions in an endemic area. *Ticks and Tick-borne Diseases*, 7(2), pp. 331–337, 2016.

[16] Slunge, D., The willingness to pay for vaccination against tick-borne encephalitis and implications for public health policy: Evidence from Sweden. *PloS One*, **10**(12), 2015.

[17] Benson, C., Watson, P., Taylor, G., Cook, P. & Hollenhorst, S., Who visits a national park and what do they get out of it?: A joint visitor cluster analysis and travel cost model for Yellowstone National Park. *Environmental Management*, **52**, pp. 917–928, 2013.

[18] Süss, J., Klaus, C., Gerstengarbe, F.W. & Werner, P.C., What makes ticks tick? Climate change, ticks, and tick-borne diseases. *Journal of Travel Medicine*, **15**(1), pp. 39–45, 2008.

[19] Centers for Disease Control and Prevention, Reported cases of Lyme disease by state or locality, 2009–2018. www.cdc.gov/lyme/stats/tables.html. Accessed on: 14 May 2020.

[20] Acadia, National Park Service. www.nps.gov/acad/index.htm. Accessed on: 14 May 2020.

[21] Scheaffer, R.L., Mendenhall, W., Ott, R.L. & Gerow, K.G., *Elementary Survey Sampling*, 7th ed., Cengage Learning: Boston, 2012.

[22] Wilkins, E. & De Urioste-Stone, S.M., Place attachment, recreational activities, and intended future visitation under climate change conditions. *Journal of Sustainable Tourism*, **26**(5), pp. 798–811, 2018.

[23] Dillman, D.A., Smyth, J.D. & Christian, L.M., *Internet, Phone, Mail, and Mixed-Mode Surveys: The Tailored Design Method*, 4th ed., Wiley: Hoboken, 2014.

[24] Zanon, D., Hall, J., Lockstone-Binney, L. & Weber, D., Development of a whole agency approach to market segmentation in parks. *Journal of Leisure Research*, **46**(5), pp. 563–592, 2014.

[25] Wilkins, E., De Urioste-Stone, S.M., Weiskittel, A. & Gabe, T., Weather sensitivity and climate change perceptions of tourists: A segmentation analysis. *Tourism Geographies*, **20**(2), pp. 273–289, 2018.

[26] Delacre, M., Lakens, D. & Leys, C., Why psychologists should by default use Welch's t-test instead of Student's t-test. *International Review of Social Psychology*, **30**(1), p. 92, 2017.

[27] Gstaettner, A.M., Lee, D. & Rodger, K., The concept of risk in nature-based tourism and recreation: A systematic literature review. *Current Issues in Tourism,* **21**(15), pp. 1784–1809, 2018.

[28] Omodior, O., Pennington-Gray, L. & Donohoe, H., Efficacy of the theory of planned behavior in predicting the intention to engage in tick-borne disease personal protective behavior amongst visitors to an outdoor recreation center. *Journal of Park and Recreation Administration*, **33**(2), pp. 37–53, 2015.
[29] Jones, E.H., Chanlongbutra, A., Wong, D., Cunningham, F. & Feldman, K.A., Knowledge, attitudes, and practices regarding Lyme disease prevention among employees, day visitors, and campers at Greenbelt Park. *Park Science*, **32**(2), pp. 46–53, 2015–2016.
[30] Mowbray, F., Amlot, R. & Rubin, G.J., Predictors of protective behaviour against ticks in the UK: A mixed methods study. *Ticks and Tick Borne Diseases,* **5**(4), pp. 392–400, 2014.
[31] Aenishaenslin, C., Bouchard, C., Koffi, J.K., Pelcat, Y. & Ogden, N.H., Evidence of rapid changes in Lyme disease awareness in Canada. *Ticks and Tick-borne Diseases*, **7**(6), pp. 1067–1074, 2016.

EVIDENCE-BASED DESIGN OF A TOURIST BROCHURE BASED ON WORSHIPING BEHAVIOR IN LONGSHAN TEMPLE, TAIWAN

MENG-CONG ZHENG & TZU-CHUN TSAI
Department of Industrial Design, National Taipei University of Technology, Taiwan

ABSTRACT

The number of visitors to Taiwan is increasing, but most of them experience it with at just a passing glance and shallow experience. The purpose of this study is to redesign the tourist brochure to provide practical information for visitors. This study was implemented based on the Longshan Temple. The first stage investigated 50 tourists' behavior in the Temple. We found the utilization rate of worshiping information was only 20%. Search time was 15% of the total visiting time; 38% of the visitors were confused because they were looking for where the worshiping objects were. Most visitors left after taking photos, unable to understand and experience the culture of the temple. In the second stage, based on the results of the visiting tasks and questionnaire, we found the "Visit ceremony", "Visit order", "guide map", and "temple history story" of the new tourist brochure should be redesigned. In the third stage, the verification results show that the correct rate of the new brochure for worshiping Longshan Temple is higher than 80%. The information about the order of visit is critical for worshiping. The new tourist brochure design effectively enhances the visitor experience.

Keywords: evidence-based design, tourist brochure, worshiping behavior, Longshan Temple.

1 INTRODUCTION

Tourism is an economic activity with relatively small impact, it can promote social and cultural progress and improve local residents' welfare [1]. Urban tourism has received little attention among studies of urban planning and sustainable tourism [2], which mostly emphasizes environmental problems [3]–[5]. Increasingly, case studies have discussed the impact of tourism on specific cities by identifying the factors affecting residents' satisfaction [6] as well as the effects of protest movements and the media [7], [8]. The tourism industry should have the potential to promote urban development, encouraging cultural exchange, and thus creating a civilized, vibrant, and safe urban environment [9]–[11].

Religious tourism is a multifaceted tourism category, and many countries regard religious culture as a key tourism asset. Taiwan has a diverse and unique religious culture. Statistics show that the total number of temples in Taiwan has reached 13,305; the popularity of religion in Taiwan is summarized in the expression "a small temple within five steps and a large temple within ten steps". At various well-known religious sites in Taiwan, tourist numbers even exceed those at natural scenic areas and amusement parks [12]. Religious tourism locations also encourage the development of the peripheral economy, including transportation, accommodation, catering, and souvenirs [13]. Shinde [14] refers to Rinschede [15] and Shinde [16]: "Religious tourism refers to contemporary patterns of visitation to places of religious importance or pilgrimage sites where visitors aim to fulfill religious needs and recreational needs". Ebadi [17] found that people visit the same site with very different motivations. In the same place, they are religious pilgrimage for some while encompassing pilgrimage or tourism for others. The value of religious tourism has been found in recent years. Jauhari and Sanjeev [18] propose that religious tourism is seen as travel with a purpose and resistant to the current economic slowdown. Melian et al. [19] propose that both religious and secular tourists gave importance to different factors inaccessibility. On the other hand,

WIT Transactions on Ecology and the Environment, Vol 248, © 2020 WIT Press
www.witpress.com, ISSN 1743-3541 (on-line)
doi:10.2495/ST200081

there is a diversity of religious tourism around the world. Sharma and Sharma [20] mention the fairs and festivals at Pithoragarh in India, have the expression of religious values. The Yanshuei Fireworks Festival in Taiwan started as a local religious ritual but is now a tourist attraction as well [21]. Clarke [22] refers to the Santa Claus Village theme park in Finland, which uses Christmas as a tourist "pull". Poranamond [23] similarly asserts in the context of sand pagodas in Thailand: "traditional religious practices have been transformed into commercialized tourist attractions" for people who may not be religious tourists. Service provision and amenities are important even for them [24], [25]. Therefore, how to improve the attraction of religious tourism locations is an important topic for the development of religious tourism.

Holbrook and Hirschman [26] believed that a tourist product covers the physical object itself, the service, and the consumer experience, through which tourists can form long-lasting memories. They believed that experience stems from tourists' pursuit of adventure and happiness. Pine et al. [27] argued that experience should be tourist-centered and that services and products should be used to create memorable activities. During the process of consumption, tourists not only invest money but also time and effort; thus, tourists also hope to obtain their desired outcome [28], which is contained within the consumer experience. Tourists care about the depth and breadth of the experience rather than the size or quantity of the product [29]. Increasingly, cities are being shaped into creative spaces to encourage the sharing of experiences between residents and tourists [30]. This includes the application of concepts such as sustainable experience design, location making, and hospitality.

Longshan Temple in Wanhua District, the National Palace Museum, and the Chiang Kai-Shek Memorial Hall are three major tourist destinations for foreign tourists visiting Taipei. Longshan Temple was selected for the present study. Constructed in 1738, it is a temple that represents the integration of Confucianism, Buddhism, and Taoism in Taiwan, and it has special rules of worshiping that differ from those of other temples. Longshan Temple currently has three types of wayfinding information for visitors: a navigation map, direction indicators, and navigation brochures. However, the existing information does not enable tourists to complete their visit smoothly and deprives them of an in-depth experience and understanding of the unique worship culture of Longshan Temple and Taiwan's temple culture. Among the three types of wayfinding information at Longshan Temple, navigation brochures are the most flexible and can carry more information, which is required given the complexity of the temple's cultural rules. Zhou [31] discovered that tourists often use navigation brochures to plan travel itineraries and visit tourist attractions, and that half of tourists refer to tourist brochures during their travel. The low cost, easy access, and portability of navigation brochures makes them an indispensable advertising media for marketing and tourism. Navigation brochures are also the most common tool for tourist sites to convey information to tourists. The main purpose of navigation brochures is to convert potential tourists into actual tourists, who then visit local attractions, through tourism marketing [32]–[34]. The purpose of this study was to design a navigation brochure to enable tourists to have a deep experience of Taiwan's temple culture at Longshan Temple, thereby reducing excessive superficial sightseeing. First, the characteristics of tourists' wayfinding behaviors at Longshan Temple were observed, and the relevant mobility problems were formulated through simulated worship experiments. Second, tourists' expectations regarding navigation brochure information were studied. Finally, the navigation brochure was proposed for verification, thereby providing a reference for future designs.

2 METHOD

Longshan Temple in Wanhua District was the observation and research site. In this study, Longshan Temple was divided into the front pavilion, front hall, center hall, main hall, rear hall, and outdoor area. Area A – part of the front pavilion – is the plaza in front of Longshan Temple and the location of celebrations during festivals. Area B – also part of the front pavilion – is the main entrance where the incense and candle shop, service center, tourist service center, and souvenir shop are located. In this area, visitors can light their incense and find the fortune stick holders, fortune stick closets, and divination blocks. Area C consists of the center hall, which is a roofless area where the incense furnace of the Guanyin is placed. Area D is the front hall, which includes the cloister, Yuantongbao Hall, the donation site, and the information for fortune stick interpretation. Area E is the rear hall, where most of the shrines are located, including the Huatuo Hall, Wenchang Dijun Hall, Mazu Hall, Guansheng Dijun Hall, and Yue Lao Hall. Area F is the outdoor area, which includes the flower beds, bathrooms, and barrier-free ramps (Fig. 1). Behavioral observation, task Analysis and questionnaire for the interview were implemented in this study. Behavioral observation is referred to as sequential action marking; this method is a type of continuous observation-recording, and its objective is to record the relationship between time and actions. Task Analysis aims to observe participants to understand the process in action, to learn more about the user how to perform a task, and achieve the user expectations [35]. The investigation was divided into three stages:

Stage I: A total of 50 tourists who speak foreign languages were observed with behavioral mapping, including where they visited, wayfinding, and sightseeing behaviors at Longshan Temple in one day. Foreign tourists were identified at the entrance of Longshan Temple randomly based on the language, then recorded the entire process by GoPro until the tourist left the temple. These behaviors were analyzed to assess the problems that tourists encounter at the temple.

Stage II: Worship task experiment. A total of 30 individuals between 20–40 years old who never visited Longshan Temple were assigned a worship task that simulated the sightseeing situation based on the complete worship process, including holding incense, lighting candles, casting divination blocks, and drawing fortune sticks. Then asked to complete a questionnaire and answer interview questions to understand the information that they referred to when performing the tasks and where they encountered difficulties to determine whether the wayfinding information at Longshan Temple was useful to worshipers. The whole experiments of stage II were implemented in two weeks.

Stage III: Design proposal and verification. The findings from stage I and II were summarized into conditions that served as the basis for brochure design. For the verification stage, eight men and eight women between 20–40 years old who never visited Longshan Temple were recruited. All participants were requested to complete the worship process based on the instructions in the navigation brochure. Upon completion, they were asked to complete a questionnaire, draw a cognitive map, and answer interview questions to evaluate their experience. Stage III was implemented in one week.

Figure 1: Five areas of Longshan Temple and behavioural mapping from non-participant observation.

3 RESULTS

3.1 Stage I: Non-participant observation on tourist behaviors

During the first stage of observation, the 50 foreign tourists' average length of stay at Longshan Temple was 24 minutes 18 seconds: among the six areas, the longest stay was in the front hall of Area B (8 minutes 12.2 seconds), followed by Areas E, A, C, D, and F. The traffic flow in Area A was chaotic because tourists would walk around to take photographs.

In Area B, tourists moved back and forth several times to locate the place to obtain incense, resulting in an overlapped traffic flow. The flow in Area D was more consistent due to the narrow passage. Most tourists in Area E walked counterclockwise from Huatuo Hall to Yue Lao Hall and then walked around Yue Lao Hall to receive their red thread, resulting in a relatively disordered flow. In addition, observation revealed that the navigation maps at Longshan Temple were placed on both sides of the wall at the Longmen entrance of Area B (front hall). The maps on the right side of the entrance are Chinese maps and were obscured by an automated teller machine. The maps on the left side of the entrance had a usage rate of only 8%. Guidance signs were installed on the pillars at the entrance to Area C from Area B and the entrance to Area E from Area D, as well as on the door to Area F from Area E; however, because of their installation height, they were easily obstructed by the crowd, resulting in a 10% usage rate. The navigation brochures, which were placed on the service counter to the left of Area B (front hall), had a usage rate of only 16%. Among the six areas, the average times for looking around, asking for directions, looking at the map, and looking at the signs were highest in Area B. When entering Area B, tourist often looked around to understand the environment. If they missed the incense and candle shop, they searched for the placed to acquire incense and thus asked for directions. The map was located in Area B (front hall), and tourists would look at the map as soon as they discovered it when entering from the Longmen entrance.

Checking the guide and obtaining information from mobile phones mostly occurred when tourists were confirming the procedures for drawing fortune sticks and casting divination blocks. In Area C – a roofless, well-lit transition area – the most common activity was looking around. Reading guidebooks was the most common behavior in Area F (Figs 1 and 2).

The average times of worshiping and nonworshiping visitors were 26 minutes and 16 seconds and 17 minutes and 21 seconds, respectively. Observations revealed that 27.3% of visitors did not worship because they could not locate their incense in Area B. One nonworshiping visitor purchased souvenirs, whereas 14 worshiping visitors purchased souvenirs. Because most of the shrines in Longshan Temple are located in Area E, this area had the second highest average sightseeing time of all areas (for both worshiping and nonworshiping visitors). Nonworshiping visitors left the temple area immediately after exiting the temple (Fig. 3). None of the 39 foreign tourists who acquired incense and worshiped the deities were able to perform a complete worship, with most visitors making mistakes in the order of halls to worshiping at. The maps and navigation brochures at Longshan Temple only indicated the order of the incense burners and did not specified the order of halls, traffic flow, or worshiping process. Furthermore, tourists hesitated at the turning points of routes and in well-lit transition areas. Because the location of the incense and candle shop was unclear, tourists would return to the area after missing the shop, resulting in a disordered traffic flow. Only approximately 20% of the tourists used the wayfinding guidance system in the temple. The average staying time of tourists at Longshan Temple was 24 minutes and 18 seconds, and wayfinding accounted for nearly 15% of this time. Of the 78% of foreign tourists who worshiped the deities, none completed the correct worship process. Tourists who worshiped the deities stayed longer than the nonworshipers. However, 27.3% of worshiping visitors forewent the worship process because they failed to locate the incense shop in Area B (front hall). Nonworshipers purchased fewer souvenirs than worshipers. These results demonstrated the numerous problems with the worship information in Longshan Temple. Of the three guiding tools in the temple (i.e., navigation maps, navigation brochures, and guidance signs), only navigation maps and brochures indicated the order of incense burners; however, the lack of other worship information resulted in a usage rate of less than 20%.

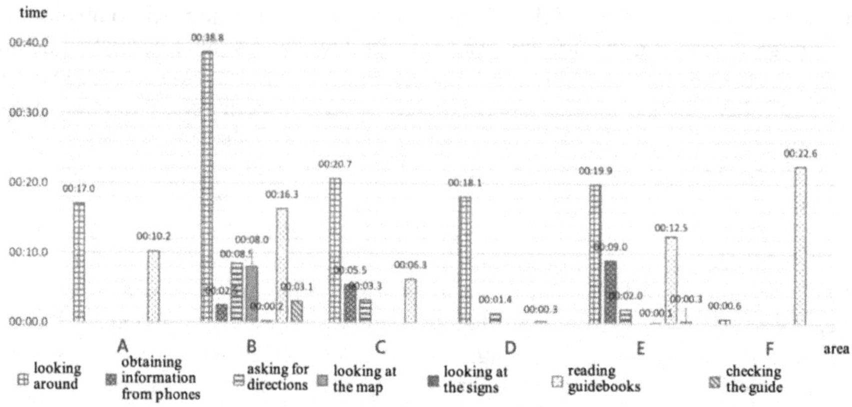

Figure 2: The average times of information searching behavior in each Area.

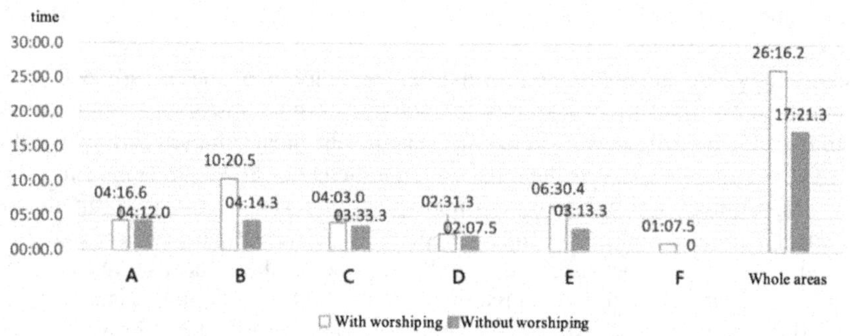

Figure 3: The average times of worshiping in each Area.

3.2 Stage II: Worship task experiment

The findings of the worship simulation experiment were similar to those of the observation in the first stage. Because the wayfinding information for worshiping in Longshan Temple was vague, the participants mostly asked others to acquire information about the worshiping tasks such as lighting candles, casting divination blocks, and drawing fortune sticks. The participants exhibited the most disordered traffic flow at Area B (front hall), and they repeatedly walked between the candle and incense shop and the incense burner to confirm the object positions. Most participants stated that the wayfinding information in the temple was not helpful for their tasks, and they obtained useful information by asking volunteers or other worshipers. They all encountered difficulties in "location" and "process". When asked about the importance of worship information, the participants ranked "the significance and process of worship with incense" as the most important, followed by "the significance and process of drawing fortune sticks", "worship order", "the significance and process of casting divination blocks", "introduction of deities", and "the significance and process of lighting candles". The participants ranked "historical stories" and "temple art" as the least important.

3.3 Stage III: Design proposals and verification

Most navigation brochures provided in temples have a folded area of less than 10.5 × 21.0 cm^2 (Cheng, 2009); their dimensions are mostly in the range 10.0–10.5 × 18.8–21.5 cm^2, which is convenient for tourists to collect and carry. However, these brochures typically do not contain introductions to the worship process at the temple and mostly focus on historical stories, temple art, and deities. The Longshan Temple navigation brochure designed in this study was assigned a dimension of 39 × 21 cm^2, with a default paper size of 8K (39 × 27 cm^2). The brochure was a five-fold map, where the width of one fold was 6.5 cm, and it complied with the design conditions that the brochure should be smaller than 10 × 21 cm^2. The outer wall color of Longshan Temple (light khaki, C: 1, M: 7, Y: 17, K: 0) served as the main background color of the brochure. The contents of the brochure were prepared on the basis of the questionnaire and interview results; they comprised "worship ceremony: light candles, worship with incense, cast divination blocks, request a red thread, and draw fortune sticks", "guidance on worship order", "floor plan", "facility configuration", "deities introduction", "temple art", and "historical stories". The preliminary layout for the brochure was based on a single fold (6.5 × 21 cm^2) as the unit. The navigation map was a three-dimensional map of Longshan Temple illustrated using 3D drawing software. Deity images were inserted into the map according to their shrine locations, and short texts were included to introduce them. Border design elements same as the front and back covers were used to distinguish the deity images from the navigation map, and arrows were incorporated into the border patterns to indicate the locations of the shrines. Finally, images of the worship ceremony were added with numbers indicating the order of worship for tourists to refer to (Fig. 4).

Figure 4: New navigation brochure of Longshan Temple.

The third stage of the experiment revealed that all 16 participants had performed worship actions such as lighting candles, worshiping with incense, casting divination blocks, and drawing fortune sticks. The task with the highest accuracy rate was casting divination blocks and requesting a red thread (100%), followed by worshiping with incense (93.75%) and drawing fortune sticks (93.75%). The correct worship order was “light candles → worship with incense → cast divination blocks → draw fortune sticks”, and 15 of the 16 participants performed with the worship in the correct order, with an accuracy rate of 93.75%. The overall traffic flow and moving sequence were more consistent than in the previous case (Fig. 5).

Analysis of worship actions when using the new brochure revealed the following findings:

- Lighting candles: The participants referred to the order of worship when lighting candles to understand their significance. Participants who referred to the information on worship ceremony and worship order accounted for 81.25%. These participants exhibited behaviors of repeatedly checking the content of the brochure.
- Worshiping with incense: Participants read the folder when walking during their worshiping process, which affected the traffic flow of other worshipers and required improvement. The participants referred to the brochure to confirm their next action when performing a worship sequence. All participants referred to the worship order.
- Casting divination blocks: Before casting the divination blocks, the participants read the worship order to identify the correct location and understand how to cast the blocks. Most participants read the brochure at fixed location. A total of 14 participants referred to the information on the worship ceremony, and nine of them referred to the worship order.
- Requesting a red thread: Most participants did not notice this information; thus, the brochure layout required further adjustment.
- Requesting fortune sticks: Fortune stick holders and divination blocks are placed in the same location. The participants knew the location of the holders if they had found the divination blocks; thus, they only read the worship order after obtaining their fortune stick number.

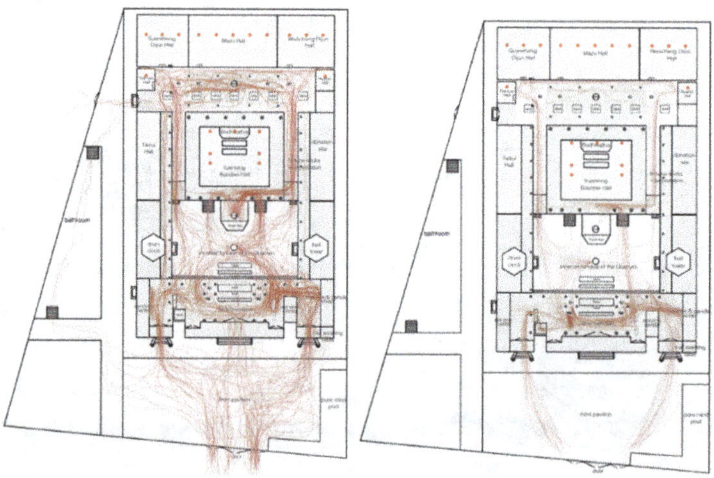

Figure 5: Traffic flow of Longshan Temple with and without new brochure.

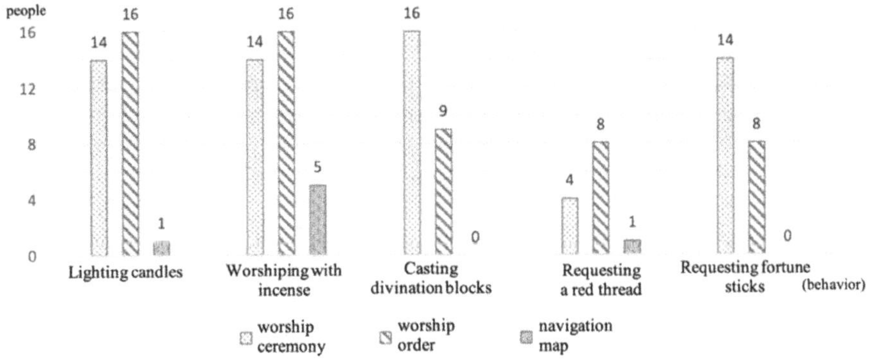

Figure 6: Utility rate of brochure information in each worshiping behavior.

All participants referred to the worship order in the brochure when lighting the candles and worshiping with incense. However, the number of participants who referred to the worship order was lower when casting the divination blocks and requesting fortune sticks; for these tasks, most participants referred to the information on the worship ceremony. This indicated that lighting candles and worshiping with incense required worshipers to move, thus increasing the importance of the floor plan and worship order. Because casting divination blocks and requesting fortune sticks were performed in the same location, the participants referred to the information on the worship ceremony after identifying the relevant location. Few participants used the navigation map; the participants stated that because the worship order had been marked on the floor plan, they were able to understand the spatial configuration of Longshan Temple. In general, the information that was referred to by most participants when lighting candles was worship order and worship ceremony (13 participants). The participants spent the longest average time reading the information on the worship ceremony. When the participants were lighting candles, the highest result for participants looking around (wlr) was observed in Area B (front hall). Participants primarily referred to the information on worship order and the worship ceremony when worshiping with incense, and they spent the longest average time on reading the worship order. When the participants were casting divination blocks, the information most often referred to was also worship order and worship ceremony; however, they spent the longest time in reading about the worship ceremony. Finally, the nine participants who requested a red thread referred to the information on worship order. The aforementioned findings demonstrate that the information within the temple should correspond to the diverse behaviors and needs of visitors (Fig. 6).

4 DISCUSSION

The observations revealed that the tourists exhibited disordered traffic flow during worshiping. During the second stage of the experiment (worship simulation experiment), 18 of the 30 participants (60%) walked back and forth; however, this decreased to 31.25% during the third stage (verification experiment). During the second stage, the average time each participant spent looking around when lighting the candles was 14.1 seconds, whereas this decreased to 4.4 seconds during the third stage.

Moreover, during the second stage, the average time each participant spent looking around when worshiping with incense was 11.6 seconds, whereas this decreased to 2.7 seconds during the third stage. The observations revealed that the participants spent more time looking around for the incense before implementing the new navigation brochure. After learning from the brochure, they were more certain that the correct worship process was lighting a candle followed by worshiping with incense. During the second stage, 25 participants performed the action of worshiping with incense first; however, only one participant did so during the third stage (Fig. 7).

During the second stage, the average time each participant spent looking around when casting divination blocks was 2.1 seconds, whereas this decreased to 1.2 seconds during the third stage. Although this difference was small, the participants were able to locate the fortune stick holder in the temple. Therefore, marking the location of the holder on the floor plan of the navigation brochure only yielded limited improvement. Similarly, during the second stage, the average time each participant spent looking around when requesting fortune sticks was 3.5 seconds, whereas this decreased to 1.7 seconds during the third stage. In both stages, the participants were able to find the location for requesting the fortune sticks.

The accuracy rates for lighting candles, worshiping with incense, casting divination blocks, requesting fortune sticks, and the overall worship process in the verification experiment showed huge improvements relative to the worship task experiment. Moreover, the accuracy rates of worship actions in the verification experiment were all higher than 80%. Reisinger and Mavondo [36] noted that tourists may feel anxious if they have trouble finding their way during sightseeing and may be less willing to return in the future. The navigation brochure designed in this study was inferred to effectively guide worship action (Fig. 8).

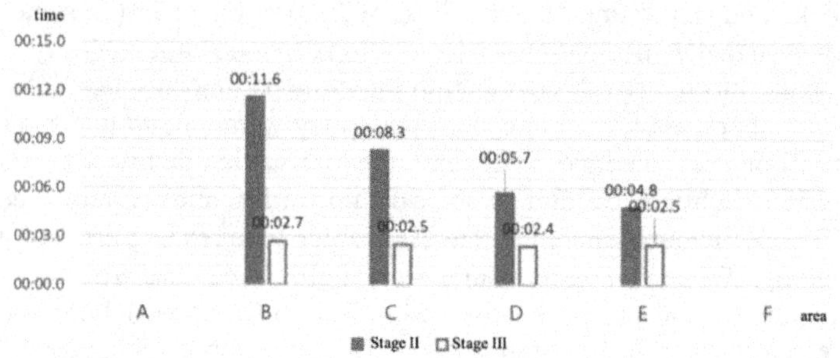

Figure 7: Average time each participant spent looking around when worshiping.

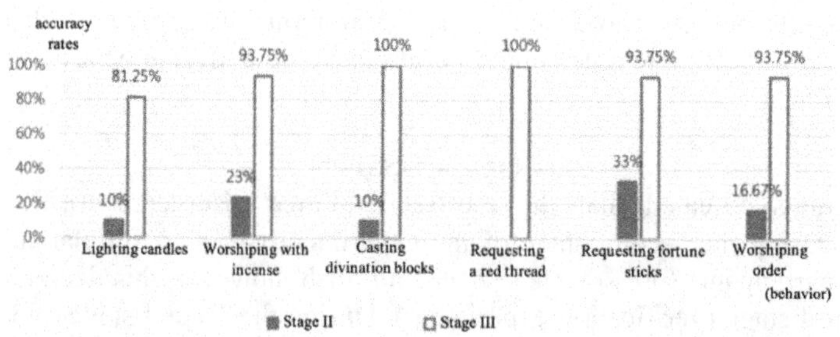

Figure 8: Accuracy rates of each worshiping behavior.

5 CONCLUSION

Observations in this study revealed that foreign tourists who tried to worship at the temple lacked understanding of the worship process because of the numerous unclear signs. They repeatedly walked back and forth, resulting in disordered traffic flow. In addition, 27.3% of the tourists did not worship because they could not locate the incense shop. This study revealed that foreign tourists' average length of stay at Longshan Temple was approximately 24 minutes and 18 seconds; 15% of this time was spent on wayfinding. In particular, wayfinding time in the front hall was higher than that in other areas, indicating that the information within the temple remains incomplete. Visitors often asked others for information on worshiping in Longshan Temple, which increased the burden on volunteers. Longshan Temple has three types of guiding objects: a navigation map, navigation brochures, and guidance signs, with usage rates of less than 20%. Surveys conducted in this study showed that tourists wanted information about the worship ceremony when visiting the temple. In the hand-drawn map survey, respondents often drew the objects mentioned in the worship order, clearly indicating that location information affected their spatial perception. Therefore, the spaces that visitors use should be considered when designing the location-related information. Circuitous tours enable tourists to know better particular areas and travel ranges [37]. It can be defined according to the moving distances, moving ranges, and traversed areas of tourists [38]. We must acknowledge the visibility of environmental features can encourage tourist familiarization with a place [39]. Therefore, site maps must both support circuitous movements and provide enough information to help tourists plan. Providing new tour information is an important factor in promoting the frequency of circuitous tour movements. Information intended for circuitous tour movements provides richness and freedom of choice, though it might not indicate the shortest travelling route. Adding directional information can enable tourists to move more smoothly during their visit. The navigation brochure provided at Longshan Temple only introduced temple-related historical stories and art; however, this study revealed that historical stories and temple art are of low importance to visitors. Thus, designers must consider whether to present the relevant information in brochures and maps or incorporate it in guidance signs. To provide appropriate information, designers must also evaluate the different behaviors of visitors to Longshan Temple, including whether they are moving continuously or worshiping at fixed locations. Experience is one of the most crucial values in religious tourism, and it enables visitors of different religions to understand the meaning and central ideas of the religious site. Information on the worship process can be provided during worship ceremonies to give visitors a profound impression when experiencing the religious actions. Participation in worshiping increased visitors' willingness to purchase souvenirs in the present study, thereby increasing profit for the tourist attraction. This study demonstrated that an improved navigation design encourages a satisfactory tourism experience. The research process and method can serve as a reference for future studies on sustainable tourism.

REFERENCES

[1] Hall, C.M., Tourism urbanization and global environmental change. *Tourism and Global Environmental Change: Ecological, Economic, Social and Political Interrelationships*, eds S. Gossling & C.M. Hall, London: Routledge, pp. 142–156, 2006.

[2] Ashworth, G. & Page, S.J., Urban tourism research: Recent progress and current paradoxes. *Tourism Management*, **32**(1), pp. 1–15, 2011.

[3] Bramwell, B., Theoretical activity in sustainable tourism research. *Annals of Tourism Research*, **54**, pp. 204–218, 2015.

[4] Buckley, R., Sustainable tourism: Research and reality. *Annals of Tourism Research*, **39**(2), pp. 528–546, 2012.
[5] Ruhanen, L., Weiler, B., Moyle, B.D. & McLennan, C.J., Trends and patterns in sustainable tourism research: A 25-year bibliometric analysis. *Journal of Sustainable Tourism*, **23**(4), pp. 517–535, 2015. https://doi.org/10.1080/09669582.2014.978790.
[6] Kuščer, K. & Mihalič, T., Residents' attitudes towards overtourism from the perspective of tourism impacts and cooperation – The case of Ljubljana. *Sustainability*, **11**(6), 1823, 2019. https://doi.org/10.3390/su11061823.
[7] Hughes, N., Tourists go home': Anti-tourism industry protest in Barcelona. *Social Movement Studies*, **17**(4), pp. 471–477, 2018.
[8] Milano, C., Overtourism, social unrest and tourismphobia. A controversial debate. *PASOS: Revista de Turismo y Patrimonio Cultural*, **16**(3), pp. 551–564, 2018.
[9] Fainstein, S.S., *The Just City*, Cornell University Press: Ithaca, 2010.
[10] Sennett, R., *The Fall of Public Man*, W.W. Norton & Company: London, 1992.
[11] Sennett, R., *Together: The Rituals, Pleasures and Politics of Cooperation,* Yale University Press: New Haven, 2013.
[12] Dahe, L., *Tourism and Cultural Festivals-Taiwan's current tourism development focus, Research on APEC Issues for Tourism and Cultural Festivals*, APEC Research Center: Taipei, pp. 14–21, 2011.
[13] Jian-Ming, C., Tourism attractiveness of religious sites in Taipei: A case study of Lungshan Temple and Hsingtien Temple. Master's thesis, National Taiwan University. https://hdl.handle.net/11296/287bm7, 2005.
[14] Shinde, K.A., Religious tourism and religious tolerance: Insights from pilgrimage sites in India. *Tourism Review*, **70**(3), pp. 179–196, 2015.
[15] Rinschede, G., Forms of religious tourism. *Annals of Tourism Research*, **19**(1), pp. 51–67, 1992.
[16] Shinde, K.A., Visiting sacred sites in India: Religious tourism or pilgrimage? *Religious Tourism and Pilgrimage Festivals Management: An International Perspective*, eds R. Raj & N.D. Morpeth, CABI: Wallingford, pp. 186–199, 2007.
[17] Ebadi, M., Typologies of the visitors at Khaled Nabi Shrine, Iran: tourists or Pilgrims? *International Journal of Culture, Tourism and Hospitality Research*, **8**(3), pp. 310–321, 2014.
[18] Jauhari, V. & Sanjeev, G.M., Managing customer experience for spiritual and cultural tourism: an overview. *Worldwide Hospitality and Tourism Themes*, **2**(5), pp. 467–476, 2010.
[19] Melian, A.G., Prats, L. & Coromina, L., The perceived value of accessibility in religious sites – do disabled and non-disabled travellers behave differently? *Tourism Review*, **71**(2), pp. 105–117, 2016.
[20] Sharma, S. & Sharma, S., Understanding Pithoragarh's cultural jewels: its fairs and festivals. *Worldwide Hospitality and Tourism Themes*, **7**(4), pp. 331–346, 2015.
[21] Cheng, T.-M., Chang, S. & Dai, Z., The mediating effect of situation involvement of novelty seeking and satisfaction for Yanshuei fireworks festival in Taiwan. *International Journal of Culture, Tourism and Hospitality Research*, **9**(2), pp. 154–167, 2015.
[22] Clarke, P., Christmas gift giving involvement. *Journal of Consumer Marketing*, **23**(5), pp. 283–291, 2006.
[23] Porananond, P., Tourism and the transformation of ritual practice with sand Pagodas in Chiang Mai, Northern Thailand. *Tourism Review*, **70**(3), pp. 165–178, 2015.

[24] Durgamohan, M., Cultural and economic significance of Hampi festival. *Worldwide Hospitality and Tourism Themes*, **7**(4), pp. 377–392, 2015.

[25] Jauhari, V., Hospitality, tourism and economic growth in India. *Worldwide Hospitality and Tourism Themes*, **1**(1), pp. 7–11, 2009.

[26] Holbrook, M.B. & Hirschman, E.C., The experience aspects of consumption Consumer fantasies, feelings, and fun. *Journal of Consumer Research*, **9**(2), pp. 132–140, 1982.

[27] Pine, B.J., Pine, J. & Gilmore, J.H., *The Experience Economy: Work Is Theatre & Every Business a Stage*, Harvard Business Press: Cambridge, MA, 1999.

[28] Mathwick, C., Malhotra, N. & Rigdon, E., Experiential value: Conceptualization, measurement and application in the catalog and Internet shopping environment. *Journal of Retailing*, **77**(1), pp. 39–56, 2001.

[29] Pei-fen, W., Study on the experience value of Watson's female skin care products. *Journal of Hospitality and Home Economics*, **5**(3), pp. 295–310, 2008.

[30] Russo, A.P. & Richards, G., *Reinventing the Local in Tourism: Producing, Consuming and Negotiating Place*, Channel View Publications: Bristol, 2016.

[31] Zhou, Z., Destination marketing: Measuring the effectiveness of brochures. *Journal of Travel & Tourism Marketing*, **6**(3–4), pp. 143–158, 1997.

[32] Gets, D. & Sailor, L., Design of destination and attraction-specific Brochures. *Journal of Travel and Tourism Marketing*, **2**(2/3), pp. 111–131, 1993.

[33] Jenkins, O., Photography and travel brochures: The circle of representation. *Tourism geographies*, **5**(3), pp. 305–328, 2003.

[34] Chang, W.-C., Rethinking resource identification and utilization: The reconstruction of indigenous ethnoecological knowledge in Fata'an Wetland, Taiwan. *Management of Environmental Quality: An International Journal*, **22**(2), pp. 187–199, 2011.

[35] Hackos, J.T. & Redish, J.C., *User and Task Analysis for Interface Design*, John Wiley & Sons, 1998.

[36] Reisinger, Y. & Mavondo, F., Travel anxiety and intentions to travel internationally: Implications of travel risk perception. *Journal of Travel Research*, **43**(3), pp. 212–225, 2005. https://doi.org/10.1177/0047287504272017.

[37] Takashi, S., On the circuit system and urban environmental formation in urban tourism. *Vol. 2: On Practical Use of Historical Heritage for Urban Tourism in Himeji*, pp. 485–488, 2005.

[38] Kaori, H., On the influence of a city map as a guide for rambling downtown beyond main streets: On the citizen's rambling activity in urban streets and their spatial cognition, pp. 213–214, 2011.

[39] Yuichi, S., "The Visual Information which encourage or Restrain Citizens" Strolling Activities in Urban Space: On the relationship of strolling activities and visual information given by the environment in downtown Kumamoto Part II. *Journal of Architecture and Planning*, **72**(614), pp. 191–197, 2007.

DISASTER INFORMATION PROVISION FOR INTERNATIONAL TOURISTS USING AN INTERPRETER APPLICATION

TAKEYASU SUZUKI
Disaster and Environmentally Sustainable Administration Research Center, University of Yamanashi, Japan

ABSTRACT

With its new tourism-focus, Japan has been encouraging an increase in foreign tourists; consequently, over the past eight years, foreign tourist numbers have slowly been growing toward the 2020 target of 40 million. However, as Japan is a natural disaster-prone country, foreign tourists are often present when natural disasters occur, at which times they attempt to find information on the internet, often without success. It has been reported that while most foreign tourists seek information from the staff at their accommodation, many small-to-medium-sized tourist facilities in Japan have few employees who can speak foreign languages. Translation AI applications are now being widely used by travelers, with the author having been involved in a collaboration with several companies to develop a sophisticated interpreter application for smartphones that can handle one-to-many and multilingual translations. Therefore, based on the Tourism Crisis Management Plan, this study examined the use of this application to provide information to foreign tourists during a simulated disaster at an amusement park. The simulation was found to be extremely effective, and revealed that foreign tourist disaster information needed to account for the international differences in legal systems, disaster experiences, natural predispositions, lifestyles and cultures, and needed to provide plain language information on: (1) what was happening, (2) the disaster progression, and (3) the actions to be taken.

Keywords: disaster response simulation, foreign tourists, interpretation app, tourism crisis management, amusement park.

1 INTRODUCTION

As Japan is a natural disaster-prone country, foreign tourists have often been present at times of natural disaster, at which times they have attempted to find information on the internet often without success. It has been reported that while most foreign tourists seek information from the staff at their accommodation, many small-to-medium-sized tourist facilities in Japan have few employees who can speak foreign languages.

The development of AI technology means that it is now possible to provide information in many languages using megaphone-type translators, small voice-based translators, or smartphone translation apps that enable two-way conversations. As these types of translators are also being used in disaster management drills in many municipalities and prefectures in Japan, they could also be effective for small- and medium-sized tourist facilities during disasters. However, these translators can only translate one language at a time and when using the applications, the two people need to be talking face to face. Therefore, as it was not possible for tourist facilities to talk to tourists in multiple languages, the author began working with several companies to develop an interpreter app that enabled one-to-many, Japanese-to-multilingual conversations, and began research on the use of this app for international tourists during disasters in Japan.

Disaster information needs to be in plain language and needs to tell people: (1) what is happening, (2) what progress is being made, and (3) the actions they need to take. Therefore, to assess the viability of the interpreter application, several fundamental disaster response simulations were conducted with the author taking the role of hotel staff, and international students playing the roles of the international tourists [1]. It was clear from the simulations

WIT Transactions on Ecology and the Environment, Vol 248, © 2020 WIT Press
www.witpress.com, ISSN 1743-3541 (on-line)
doi:10.2495/ST200091

that it was also necessary to provide disaster information that encouraged international tourists to take action within the frameworks of their own cultures, disaster experiences, and legal systems. Overall, however, it was shown from the simulations that using an interpreter app for disaster response was effective for use in tourist facilities such as hotels.

Therefore, this study conducts a further disaster response simulation on international tourists using the interpreter app for a volcanic eruption disaster simulation at an amusement park with the actual employees. The purpose of this study is to verify the validity of disaster information provision methods and the ability to meet the three conditions: (1) what is happening, (2) what progress is being made, and (3) the actions they need to take: and were related to the tourists disaster experiences, natural predisposition, culture, customs and lifestyles, and legal system differences.

The disaster response simulation is conducted at the Fuji-Q Highland amusement park in Yamanashi Prefecture, Japan. Based on the simulation results, then, the important issues for the provision of disaster information to international tourists will be elucidated.

2 DISASTER INFORMATION BY INTERNATIONAL VISITORS

The Japanese archipelago has steep terrain and a fragile geology and despite its area being only 0.25% of the world, has about 7% of all active volcanoes, experiences about 20% of the world's magnitude 5 or greater earthquakes, and is subjected to one of more typhoons annually. However, due to climate change effects, the number of short-term heavy rain events has increased and the number of annual typhoon landings has doubled from 10 years ago to 4–6 times a year, with the size and scale of the typhoons expanding. Therefore, Japan suffers frequently from many natural events (hazards), such as earthquakes, tsunamis, heavy rain, heavy snowfall, typhoons, high waves and storm surges, landslides, slope failures, debris flows, and volcanic eruptions.

The Japanese government campaign to attract international tourists to Japan has significantly increased international tourism in Japan, with 31.88 million international tourists visiting in 2019, which was a significant increase on the 31.19 million visitors in 2018 and indicated the steady growth toward the 40 million target in 2020. However, as the number of tourists grows, the chance of these tourists encountering a natural disaster also increases. However, it has been reported that international visitors find it difficult to gather information from Japanese TV, radio or the internet on disaster situations or traffic conditions, especially during disasters [2]–[5]. Many international tourists are unaware of natural events such as earthquakes or typhoons as they have not experienced them before and many are very confused and distressed as they do not understand the local language and they do not know the names of the places and facilities [6].

It has been found that many international tourists seek information from the employees at their accommodation during disaster [2]–[5]. Fig. 1(a) shows the results of a questionnaire conducted by the Survey Research Center for foreign tourists on the evacuation and travel information sources used during the 2016 Kumamoto earthquake, and Fig. 1(b) shows the travel information sources during the 2018 Hokkaido Eastern Iburi earthquake, from which it can be seen that "accommodation employees" was ranked second in Fig. 1(a) and first in Fig. 1(b). In the hotel lobby, foreign tourists could ask the staff several times and get a satisfactory answer. However, in small- and medium-sized accommodation facilities, few employees have the ability to speak another language besides Japanese. Further, the tourists' home country websites ranked higher than either Japanese TV or radio, which indicated that international tourists were getting their information from their own media while in the affected areas.

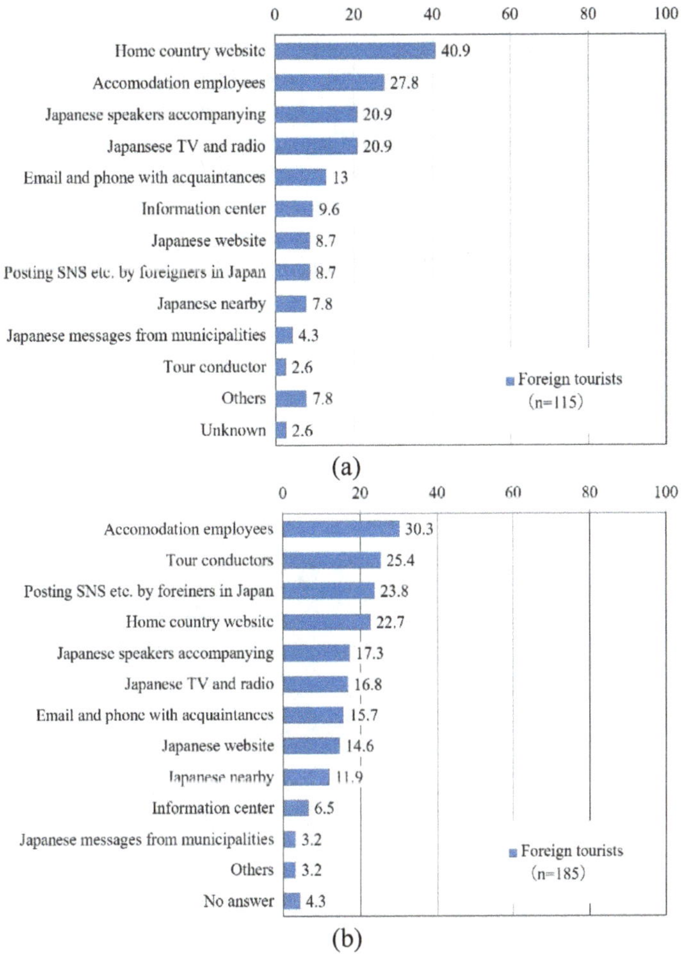

Figure 1: Necessary information for evacuation and travel during disasters. (a) The 2016 Kumamoto earthquake *(Source: Survey Research Center, 2016 [2])*; (b) The 2018 Hokkaido Eastern Iburi earthquake *(Source: Survey Research Center [4])*.

3 MULTILINGUAL INTERPRETER APP

Small- and medium-sized tourist facilities have few staff and even fewer who can speak another language besides Japanese, which makes it difficult for them to individually respond to inquiries by foreign tourists. As international visitors come from many countries, there is a need for information to be available in many different languages such as Chinese, Korean, English, Thai, Vietnamese, and Indonesian. Therefore, the author developed the idea of an interpreter app to provide information in many languages using a megaphone-type translator or a voice based small translator and a translation app that enables two-way conversations, which was consequently developed and implemented not only for normal guidance but also for disaster response in tourist facilities and for disaster management drills in municipalities, prefectures or temples and shrines with many tourists. Therefore, it is expected that this application could be used to provide tourist information during disasters in small- and medium-sized tourist facilities. However, when using these translators, it is necessary to select one language from which to translate, and in conversations using these translators or translation apps, two people have to talk face to face. Therefore, the translators and app do not allow tourist facility staff members to talk to multiple tourists in multiple languages.

For tourist facility employees to provide disaster information to multiple, multinational tourists simultaneously and to be able to answer tourist questions individually, an interpreter app that can translate from one-to-many and from Japanese-to-multiple languages is needed, which could also be used for tourist facility and guide commentaries for tourists during normal times. In this research, the "SmaLingual Pro" interpreter app, which was jointly developed by the author and Smart Culture Gateway Co., Ltd. and other companies, which has these functions, was selected (Fig. 2). This application is a paid service that provides simultaneous multilingual interpretation using a multilingual translation engine developed by the National Institute of Information and Communications Technology in Japan. The application can be downloaded from the Apple Store or Google Play, and with simple settings, it is possible to speak to multiple people in multiple languages with the same convenience as using a regular smartphone.

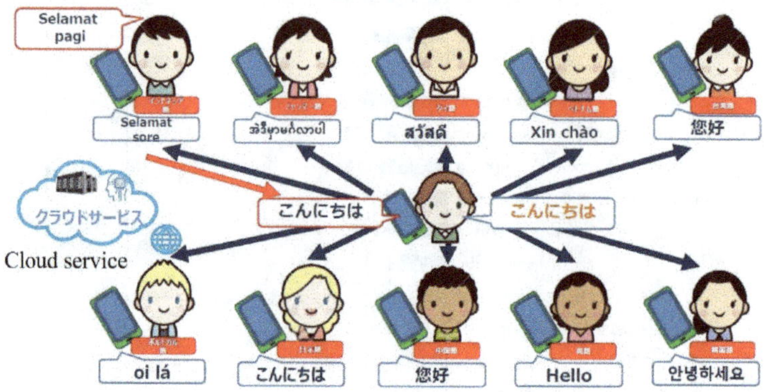

Figure 2: Schematic diagram to show features of the interpreter app. *(Source: Smart Culture Gateway Co., Ltd. [7].)*

4 CURRENT STATUS OF VOLCANIC DISASTERS AND COUNTERMEASURES

The volcanic island Whakaari/White Island in New Zealand's north-eastern Bay of Plenty region erupted on December 9, 2019. There were 47 people on the island at the time: twenty-four Australians, nine Americans, five New Zealanders, four Germans, two Chinese, two Britons, and one Malaysian. Twenty-one people were killed, including two who were missing and declared dead, and a further twenty-six suffered injuries, many of which were severe burns. Experts identified the event as a phreatic eruption, which launched rock and ash into the air.

The volcano had been showing signs of unrest for several weeks before the 2019 eruption. In October 2019, volcanic tremors and Sulphur dioxide gas were at their highest levels since 2016, indicating that an eruption was imminent, and on 18 November, the volcano was rated at a Volcanic Alert Level 2, indicating "moderate to heightened volcanic unrest". Directly after the eruption, the Volcanic Alert Level for the island was raised to a 4, but decreased by 16:30 on the same day to a level 3 [8].

A volcanologist claimed that the incident was "a disaster waiting to happen" and commented that the island was too dangerous to allow the daily tour group visits. However, even though there was the potential for eruption hazards to occur at Alert Level 2, the tours to the island were not restricted and tourists continued to regularly visit the island, primarily through White Island Tours.

In Japan, a similar phreatic eruption occurred at Mt. Ontake in September 2014, killing 58 people with a further five people missing. The Japan Meteorological Agency had observed an increase in volcanic earthquakes about two weeks before the phreatic explosion and called for caution; however, they did not raise the Volcanic Alert Level from 1, because the number of earthquakes decreased [9]. The Japan Meteorological Agency also did not give any further commentary information about the volcanic situation before raising the Alert Level. Only ten minutes before the eruption, volcanic tremors were observed due to magma ascending. Because the eruption occurred during the daytime during a holiday period, many climbers near the summit were affected.

The Japanese Meteorological Agency has an established system for setting and announcing volcanic eruption alert levels based on observation, after which the municipalities issue evacuation information based on the Alert Level, which is an index that categorizes the areas that needs alerting and the measures that need to be taken by disaster management agencies and residents based on five volcanic activity stages. The Volcanic Alert Levels in Japan, which are different from those in New Zealand [10], are applied to 41 volcanoes of the 111 active volcanoes in Japan, which are defined as those that have erupted within the last 10,000 years or those with active fumarolic activity.

Mt. Fuji is the largest active volcano in Japan, 70 craters of which, mainly at the foot of the mountain, have erupted in the past. Therefore, the Volcanic Alert Level can be raised from level 1 (active volcano) to level 3 (restriction to enter the mountain area) without going through level 2 (restriction to enter 1 km around a crater) only at Mt. Fuji. The volcanic disaster management system at Mt. Fuji requires different evacuation information to be issued in each different area based on crater location and the eruption status. Before the Meteorological Agency raises the Volcanic Alert Level from Level 1 to Level 3, it publishes volcanic observation information. However, to date, no commentary information has ever been released for Mt. Fuji since the Volcanic Alert Level announcement system was established, which means that none of the municipalities and tourist facilities at the foot of Mt. Fuji have not yet established systems to specifically respond when the volcanic observation information is announced.

5 DISASTER RESPONSE SIMULATION IN JAPANESE AMUSEMENT PARK

5.1 Need for disaster response simulations

Fundamental disaster response simulations had been previously conducted, in which the author played the role of the hotelier and international students played the roles of the tourists, from which it was confirmed that disaster information needed to be provided in plain language on: (1) what was happening, (2) the disaster progress, and (3) the subsequent actions to be taken [1]. In addition to these three conditions, the simulations also found that it was necessary to provide disaster information that encouraged international tourists to take action by accounting for their different cultures, disaster experiences, and legal systems. As the author was not familiar with hotel operations but is an expert in disaster management, it was necessary to verify the validity of the results from the fundamental disaster response simulations in an actual tourist facility.

Therefore, the author planned a disaster response simulation using the interpreter app with international tourists and the Japanese employees at a Japanese amusement park. Fig. 3 illustrates the Japanese map and the location of the Fuji-Q Highland amusement park in Fujiyoshida City, Yamanashi Prefecture, a city famous for Mt. Fuji sightseeing located at the northern foot of Mt. Fuji. Approximately 2.3 million people annually visit the tourist facilities: the Fuji-Q Highland amusement park, hotels, pleasure boats, and ropeways: operated by the Fujikyukou Group. At the meeting with representatives of the Fujikyuko Co., Ltd., the time, place, and the two participating attractions staff were selected. The two staff did not receive any special training on the measures to be taken in the case of volcanic eruptions.

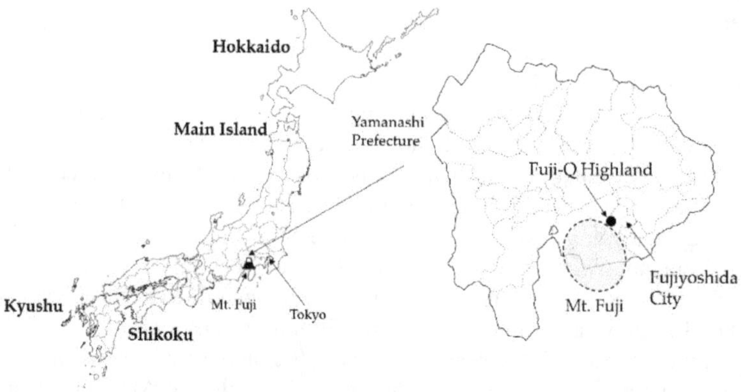

Figure 3: Japanese Archipelago and location of Fuji-Q Highland.

5.2 Simulation scenario

A simulation was conducted in which amusement park staff used the interpreter app to converse with international tourists about the volcanic situation. The simulation participants were the two staff from Fuji-Q Highland, two Chinese, and two Vietnamese. They were divided into two groups, one staff, one Chinese and one Vietnamese: with the two different groups conducting different simulation cases: direct translation of the announcement from the Japan Meteorological Agency and ingenious explanation in plain language. The simulation was based on an international tourist disaster response for a Volcanic Alert Level 3 announcement.

The simulation was conducted from 13:00 on December 20, 2019. One hour before the simulation, the Fuji-Q Highland staff participating in the simulation convened in a meeting room, where they downloaded the interpreter app onto their own smartphones and practiced how to use it. Then, the simulation preconditions shown in Table 1 were orally explained and also given in printed materials. The author also instructed each staff member about the simulation cases they were to participate in.

The simulation conditions were as follows. As shown in Table 1, two information commentaries on the situation at the volcano were released by the Japan Meteorological Agency two days before the simulation, and the staff members were informed that Fujiyoshida City had contacted Fuji-Q Highland with the details shown in Table 1. As the amusement park location was not included in the designated primary evacuation area, the

Table 1: Messages transmitted to the amusement park staff.

No.	Messages transmitted to the staff
1	On December 18, the Japan Meteorological Agency released commentary on the situation of the volcano, "We have observed volcanic earthquakes".
2	At 12:00 on December 19, the Japan Meteorological Agency announced information on the situation of volcano, stating that "The number of volcanic seismic observations is rapidly increasing" Immediately after the announcement The Volcano Disaster Management Council of Mt. Fuji was held, and the local governments participating in the council were alerted. As a result, the following contact was also made from Fujiyoshida City to Fuji-Q Highland. No evacuation information has yet been issued by the city. "Mt. Fuji's volcanic activity is somewhat intense. The number of earthquakes has increased rapidly beneath Mt. Fuji. Please be aware of future information".
3	The Japan Meteorological Agency raised the Volcanic Alert Level to 3 at 12:00 on December 20 due to the observation of an inflation of the volcanic edifice at the northern foot of Mt. Fuji.
4	As this announcement has been broadcast on TV news, to evacuate visitors from the northern foot of Mt. Fuji, Fuji-Q Highland decided to close the park in the afternoon, and decided to transport those guests without their own transportation such as private cars or sightseeing buses to JR Otsuki Railway Station using Fuji-Q's buses.
5	The amusement park decided to take special measures to guide. international tourists to the third entrance and preferentially transport them by bus to JR Otsuki Railway Station.
6	Fuji-Q Highland Resort Hotel decided to secure accommodation in Tokyo for guests planning to stay on the 20th, and to take them to the accommodation by bus. The hotel is currently making these arrangements, which are not yet finalized.

amusement park was open as usual until the day before. It had been reported on TV that the Japan Meteorological Agency had raised the Volcanic Alert Level to 3 at 12:00; however, based on its own disaster management plan, Fuji-Q Highland decided to close the park in the afternoon and to evacuate the visitors to outside the north foot of Mt. Fuji. Because of prior consultation between the author and Fuji-Q Highland, the Fuji-Q Highland decided to take special measures for the international tourists to lead them to a parking lot away from the usual entrance and then preferentially transport them by bus. For those tourists who are also guests at the Fuji-Q Highland Resort Hotel, which is adjacent to the amusement park, Fuji-Q Highland decided to secure special accommodation in Tokyo and to move them to the accommodation by bus.

Of the two Chinese and two Vietnamese participating in the simulation, three were international students at University of Yamanashi and one was the brother of a Vietnamese international student. These four international participants had learned how to use the interpreter app in advance.

The simulation conditions were described as follows: (1) you came to Japan with your friends for sightseeing, not group travel; (2) as this is your first visit to Japan, you did not understand the conditions in Japan and could not speak Japanese; (3) you have no knowledge of volcanic eruptions, as you have only watched images on TV; (4) you have been staying at Fuji-Q Highland Resort Hotel since yesterday and came to Fuji-Q Highland to have fun. Yesterday, you got off the Fuji-Q Kanko limousine bus in front of the Fuji-Q Highland Resort Hotel; (5) you visited Arakura Fuji Asama Shrine this morning, and you are going to have fun at Fuji-Q Highland, stay at the hotel, and tomorrow afternoon, take a limousine bus from the hotel to Narita Airport and return home; (6) however, a staff member of Fuji-Q Highland explains using the interpreter app (SmaLingual Pro) that the park is to be closed; however, because you are not group tour visitors, your transportation is limited to public transport.

The simulation was conducted in front of the first or main entrance of the amusement park. For simulation convenience, the four international simulation participants needed to understand that this entrance was just a few minutes' walk from the Fuji-Q Highland Resort Hotel. Therefore, the author showed them the positional relationship using a map. As the disaster information to be given by the amusement park staff was different in the two cases: direct translation and ingenious explanation, the international participants were divided into two different groups and the conversation conducted using the interpreter app.

5.3 Messages from the amusement park staff

The Volcanic Alert Level 3 information from the Japan Meteorological Agency was conveyed in two different ways based on description No. 1 in Table 2: direct translation and ingenious explanation. Then in both cases, the international tourists were given messages Nos. 2 to 4 in Table 2 by the staff.

For the international tourists accommodated at the Fuji-Q Highland Resort Hotel, the author proposed to Fujikyu Corporation that these measures be incorporated into the simulation, which was accepted. In particular, for the ingenious explanation, preferential treatment was given to guests at the Fuji-Q Highland Resort Hotel, as shown in Table 3.

Table 2: Messages given by the staff to the international tourists.

No.	Messages from staff of the amusement park	
Case	Direct translation	Ingenious explanation
1	Noticeable earthquakes have occurred and an inflation of the volcanic edifice is observed at the foot of Mt. Fuji. The Japan Meteorological Agency raised the Volcanic Alert Level up to 3. Mountain climbing on Mt. Fuji has been already prohibited.	Several signs of the possibility of an eruption were observed on Mt. Fuji, and the Japan Meteorological Agency announced that the possibility of an eruption was high. Mountain climbing on Mt. Fuji has already been prohibited. Nobody knows at this time whether Mt. Fuji will erupt or not, and when, where and how the eruption will occur.
2	For your safety, we have decided to close Fuji-Q Highland at this stage and ask all of you to evacuate to outside the northern foot of Mt. Fuji.	
3	Please follow the instructions from the staff.	
4	For foreign tourists who do not have a car or bus, please gather here. We will take you to JR Otsuki Railway Station by bus. I will guide you to the parking lot where buses will wait. Do you have any questions?	

Table 3: Preferential treatment for guests at the Fiji-Q Highland Resort Hotel.

No.	Special message for tourists staying at the Fuji-Q Highland Resort Hotel
1	For the guests staying at the Fuji-Q Highland Resort Hotel today, we are chartering a bus, which is in front of the hotel.
2	The hotel is secure. Please be assured that the bus will leave after you have prepared your luggage.
3	The Fuji-Q Highland Resort Hotel will also arrange a hotel in Tokyo.
4	The arranged bus will take you to the hotel where you will stay today.
5	We will not refund today's accommodation fee, but we will pay for your hotel accommodation in Tokyo.
6	The hotel staff will be happy to assist you.

5.4 Results of the simulation

Fig. 4 shows photographs of the simulation. Fig. 4(a) shows the staff member at Fuji-Q Highland transmitting information using the smartphone in the direct translation case, and Fig. 4(b) shows the case of ingenious explanation, in which the staff member calls the two tourists staying at the nearby Fuji-Q Highland Resort Hotel and explains the evacuation procedure by bus from the hotel.

(a) (b)

Figure 4: Photograph of disaster response experiment in the amusement park. (a) Case of direct translation; (b) Case of ingenious explanation.

Table 4 summarizes the questions and remarks from the tourists and the responses of the staff at the amusement park after the staff transmitted the messages shown in Table 2 for the information communicated through direct translation. In the speaker's column in Table 4, S1 indicates the amusement park staff, C1 indicates a Chinese tourist, V1 indicates the Vietnamese tourist, A was a question about what was happening, B was a question of how it would evolve, C was a question of what actions to take, D was a query about the jargon, E was a request from a tourist at the hotel, and F was a complaint about the hotel by a tourist.

Table 5 summarizes the questions and remarks from the international tourists and the staff responses after the staff transmitted the message shown in Table 2 for the case communicated by ingenious explanation. In the speaker's column in Table 5, S2 indicates the amusement park staff, C2 indicates a Chinese tourist, and V2 indicates a Vietnamese tourist.

The Chinese tourist said she wanted to return to the Fuji-Q Highland Resort Hotel to pick up her luggage, but the intention was not well communicated to the staff at the amusement

Table 4: Conversations between tourists and the staff in the direct translation case.

Speaker	Content of conversation	Classification
C1	Is it dangerous when the volume of the mountain increases?	D
V1	I stay at a nearby hotel. There is no car to return.	C
V1	Is the volcano erupting from now?	B
C1	What is the volcanic eruption situation?	A
V1	Please tell us about the current dangerous situation.	A
S1	I do not know.	
C1	We have to travel by bus don't we, because it is dangerous?	C
S1	Yes, please move right away.	
C1	Where are we going now?	C
S2	We are going to the parking area where the buses for evacuation park.	
V1	Is it dangerous here? Do I have to get on the bus?	C
C1	Is it far from here?	C
S1	It will take about 5 minutes on foot.	
V1	What do we do after getting on the bus? Where does the bus go?	C
C1	Is it safe to return to the hotel?	B
V1	Can I stay here for sightseeing?	C
C1	Why is Fuji-Q Highland unsafe while there is a safe place only 5 minutes by walk?	F
S1	I will guide you. Please listen.	

Table 5: Conversations between tourists and the staff in the ingenious explanation case.

Speaker	Content of conversation	Classification
C2	I need to return to the hotel to get my luggage.	C
C2	I am going back to the hotel.	C
V2	Please tell me the directions to the bus stop.	C
S2	I will guide you as soon as all guests have gathered.	
V2	I am going back to the hotel.	C
S2	Which hotel are you staying at?	
V2	I am going to the hotel in Tokyo.	C
S2	We will take you to JR Otsuki Railway Station by bus to return to your hotel in Tokyo.	
V2	I understand.	
S2	Thank you very much. Please wait a little more.	
V2	I understand.	
S2	Are there any guests staying at Fuji-Q Highland Resort Hotel? Please raise your hand if you are staying at Fuji-Q Highland Resort Hotel.	
	C2 and V2 raised their hands	
S2	The employee verbally explained what was described in Table 5.	
S2	Do you understand?	
V2	Yes, I do.	

park, and the Vietnamese tourist indicated his willingness to follow the staff's proposal to be evacuated and leave his luggage at the hotel. He could take the bus from the hotel where he was staying, while the staff member sent a message with the content shown in Table 3.

As described above, even though only five dialogues classified as C regarding the actions to be taken by the tourists were mentioned, it was found that the description of the volcano situation by the amusement park staff was understood by the two tourists. If the hotel prepares in advance to respond to guests at the Fuji-Q Highland Resort Hotel, as shown in Tables 2 and 3, the responses can be implemented more smoothly.

Fig. 5 shows a comparison of the results between the two cases in which the questions were classified into six types in the horizontal axis, with the vertical axis indicating the number of questions. In this figure, the question and remark contents from the tourists shown on the horizontal axis were classified into six types: what is happening now, how it will progress, what actions should be taken, easy-to-understand expressions, request for support, and complaint or dissatisfaction with staff responses. As shown in the figure, when the information announced by the Japan Meteorological Agency case was transmitted to the international tourists using an ingenious explanation, the number of questions was reduced to five from the thirteen for the direct translation case, and the question content was focused on the actions to be taken. Therefore, it was shown that it was more effective to explain the messages to the tourists using ingenious explanations.

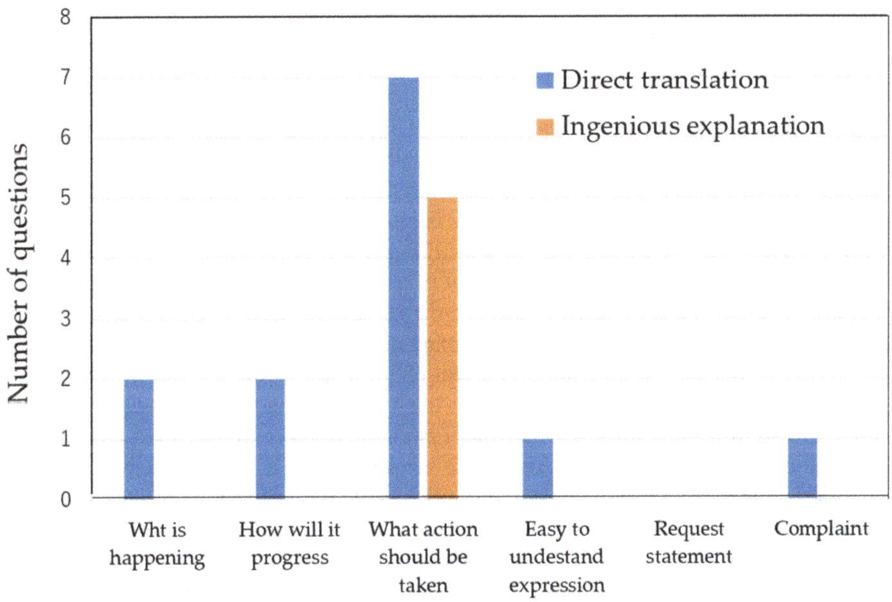

Figure 5: Comparison the results of experiment between two cases.

Despite the same evacuation guidance being given by the staff after the volcano information, there were significant differences found between the two cases. The reason for these differences was because the international tourists did not understand the technical term translations, such as "the occurrence of a felt earthquake", "inflation of the volcanic edifice", and "Volcanic Alert Level 3", as the regions from which the four international tourists were from had no natural predisposition to volcanic eruptions and therefore, they had not received disaster education on volcanic eruptions. Neither China nor Vietnam has a system for volcanic alert levels, and Vietnam has no volcanoes; therefore, the legal systems for evacuation vary from country to country. The amusement park staff who participated in the simulation had also had no prior education or training on volcanoes and volcanic eruption disaster responses, and it had been confirmed before the simulation that they also did not know the meaning of the specific jargon used by the Japan Meteorological Agency.

5.5 Discussions

Tourist facilities that do not have any expertise on volcanoes often do not know what is happening and the progress of the disaster. Therefore, municipalities need to access information immediately from specialized agencies such as the Japan Meteorological Agency and transmit it to tourist facilities in plain easy-to-understand language. The simulation in this paper verified that it is necessary to provide information to international tourists that satisfy the above three conditions.

When a Volcanic Alert Level 3 is issued, municipalities such as Fujiyoshida City are required to issue evacuation preparation information that calls for evacuation assistance for people who need evacuation assistance, but they are not legally required to issue evacuation advisory information to healthy populations. As tourists are people who need assistance during such evacuations, the author proposed closing the park and evacuating all guests outside the northern foot of Mt. Fuji. Needless to say, tourism crisis management can minimize the damage caused by disasters and allow for the resumption of the original tourism business as soon as possible after the end of the disaster [11], [12]; therefore, this simulation suggested that there was a need to have readily available tourism crisis management plans.

The amusement park staff who participated in the simulation had not had any prior education or training on volcanoes or volcanic eruption disaster response, and therefore had no knowledge of the meaning of jargonized language issued by the Japan Meteorological Agency. The one-to-many, Japanese-to-multilingual interpreter app was proven to be a powerful tool for tourist facilities in the event of a disaster. However, to make effective use of this app in an actual disaster, simulations/drills such as the one described in this paper should be conducted and disaster response manuals prepared and verified through training. Regular disaster response and manual verification simulations/drills using the interpretation application could be powerful preparation for tourist facility staff and management.

Since the simulation was conducted in front of the main entrance of Fuji-Q Highland, the volume of audio guidance broadcast from the speakers was large, and the noise when a large roller coaster in the park passed was also loud. When there were these loud sound sources around, participants had to stop talking on the smartphone. However, compared to about a few years ago, in recent years the voice recognition technology of smartphone microphones and software has been significantly advanced, and it has been confirmed that unless the wind is strong enough to be able to talk, the sound of the wind will not be a big obstacle to voice recognition.

In noisy situations, participants could pause the conversation and wait for it to be quiet. However, the accent, poor pronunciation, and habits of speakers have troubled the voice recognition function of smartphones, and as a result, have sometimes resulted in mistranslation. The staff of Fuji-Q Highland who participated in the simulation spoke not with the standard language, but with accents and speaking styles unique to the northern foot of Mt. Fuji, which seemed to reduce the accuracy of translation. Despite the hope that the translation engine will be improved, employees of hospitality industry need to be trained to speak slowly according to the disaster response manual with the correct translatable accents.

In order for international tourists to take appropriate evacuation actions according to the information, it is important for tourism and hospitality industry employees, first of all, to remain international tourists calm. Even if they send the same message, it should be received differently depending on the nationality, cultural, religious background and so on. In the simulation, the emphasis was placed on properly communicating the unknown natural threat of volcanic eruptions to international visitors, leading them to evacuate in the area strange to them. Furthermore, it is necessary to devise ways to provide information to calm such international tourists, and this will be a future issue to be studied.

Both of the two Chinese who participated in the simulation spoke Mandarin, the standard language. Therefore, a relatively appropriate translation was performed by the interpreter app. However, in the fundamental experiments conducted previously involving 11 Chinese students [1], there were many cases in which correct translation was not performed due to dialects and dullness. Therefore, the experiment involved only five students who could speak relatively standard languages. On the contrary, one of the two Vietnamese who participated in the simulation spoke the standard language, but the other one had an accent in the language, so the app was often mistranslated. Thus, the translation engine still needs improvement and expansion. Furthermore, proper nouns representing the names of facilities such as Fuji-Q Highland and Fuji-Q Highland Resort Hotel and place names such as Otsuki need to be registered in a dictionary in advance.

6 CONCLUSIONS

This paper used an innovative interpreter app in a disaster response simulation targeting volcanic eruption disasters for international tourists at an amusement park with the participation of actual staff, from which the following conclusions were derived:

1) The provision of disaster information to international tourists using ingenious explanations about (1) what is happening, (2) how it is progressing, and (3) the actions to be taken were more effective. Only five questions focused on what actions should be taken were asked by the international tourists in ingenious explanation case compared to the thirteen questions asked when the Japan Meteorological information was directly translated.
2) Neither China nor Vietnam has a system for volcanic alert levels, and Vietnam has no volcanoes; therefore, it is necessary to devise ways to provide information taking the legal systems, disaster experiences, natural predispositions in other countries. It is also necessary to remain international tourists calm during disasters in order to take appropriate evacuation actions according to the information. The way of reacting is different according to nationality, cultural background. Thus, how to calm international tourists is a future issue to be studied.
3) It is imperative that municipalities obtain information from specialized disaster management organizations such as the Japan Meteorological Agency as early as possible and transmit this information to tourist facilities using ingenious and plain language.
4) The simulation suggested that there was a need for a comprehensive tourism crisis management plan.
5) Simulations/drills at tourism facilities using the interpretation application could prepare staff for disasters and successfully verify the validity of disaster response manuals.

REFERENCES

[1] Suzuki, T. & Yang, J., Experimental study on disaster response to foreign tourists using the interpreter software. *Proceedings of the 17th International Conference on Earthquake Engineering*, Sendai, Japan, 2020.

[2] Survey Research Center, Survey on evacuation behavior of foreign tourists visiting Japan during the Kumamoto earthquake 2016. www.surece.co.jp/research/1782/. Accessed on: 14 Feb. 2020.

[3] Survey Research Center, Survey on evacuation behavior of foreign tourists visiting Japan during the Osaka Northern Earthquake. www.surece.co.jp/research/2441/. Accessed on: 14 Feb. 2020.

[4] Survey Research Center, Survey on evacuation behavior of foreign tourists visiting Japan during the 2018 Hokkaido Northern Iburi earthquake. www.surece.co.jp/research2491/. Accessed on: 14 Feb. 2020.

[5] Survey Research Center, Survey on evacuation behavior of foreign tourists visiting Japan during the 2019 off Yamagata Prefecture earthquake.

[6] Vargas-Sanchez, A., Crisis situation in tourist destinations: how can they be managed?. Enlightening Tourism. *A Pathmaking Journal*, **8**(4), pp. 47–69, 2018. www.surece.co.jp/research2968/. Accessed on: 14 Feb. 2020.

[7] Smart Culture Gateway Co, Ltd., Brochure for SmaLingual Pro 2019.

[8] Wikipedia, Whakaari/White Island eruption 2019. https://en.wikipedia.org/wiki/2019_Whakaari_/_White_Island_eruption. Accessed on: 14 Feb. 2021.

[9] Suzuki, T., *Wisdom, Techniques and Mechanisms for Protecting Lives from Catastrophes*, ITSC Ltd.: Shizuoka, 2014.

[10] Potter, S.H., Jolly, G.E., Neall, V.E., Johnston, D.M. & Scott, B.J., Communicating the status of volcanic activity: revising New Zealand's volcanic alert level system. *Journal of Applied Volcanology*, **3**(13), pp. 1–16, 2014.

[11] Richie, B.W., *Crisis and Disaster Management for Tourism (Aspects of Tourism)*, Channel View Books: Bristol, 2009.

[12] Barent, N.M., The Need to be prepared: Disaster management in the hospitality industry. *Journal of Business and Hotel Management*, **1**(2), 2012. DOI: 10.4172/2324-9129.1000101.

GETTING RESPONSIBLE TRAVELLERS ON BOARD

GIULIO PATTANARO
Independent Researcher

ABSTRACT
In times of growing global concern over the effects of climate change, railways could potentially provide the market with viable and valuable options for more environmentally-friendly responsible tourism practices. Traditional train companies themselves appear to be increasingly aware of the immense tourism potential of their sector; this is evidenced by their offer of additional – and often complementary – tourist-oriented services such as car rental bargains or combined transport–accommodation packages on top of conventional train tickets. The present paper aims to examine the marketing tools and initiatives that train companies have conceived and developed to target and attract responsible travellers. It reviews the most relevant literature on rail tourism, with a focus on both the main characteristics of rail tourists and the link between rail tourism and sustainability. The literature review will be followed by the analysis of the tools and initiatives being implemented by three national train operating companies – Société Nationale des Chemins de Fer (SNCF), Swiss Federal Railways (SBB-CFF-FFS) and Trenitalia in France, Switzerland and Italy respectively. The case study analysis is based on the inputs collected from each of the three operator's corporate sustainability strategy, business-to-consumer (B2C) website, and digital or printed promotional material provided to potential or actual passengers. The findings relative to each train operator will eventually be compared and some conclusions on the main components and outreach of the strategies and initiatives put in place by train operators to attract responsible tourists will be drawn. The paper will conclude by suggesting further research on the topic and the analysis of additional case studies.
Keywords: rail transport, rail tourism, responsible traveller, train companies, sustainable mobility.

1 INTRODUCTION

To reach a given tourism destination by train may be particularly attractive for those travellers who prefer environmentally-friendly means of transport.

For their part, traditional train companies – many of which used to be state-owned companies operating in a monopolistic market – have become more and more aware of the tourism potential of their product. This is why, next to conventional train tickets, they have come to offer additional and complementary tourist-oriented services like car rental and hotel booking.

This paper analyses the strategies and marketing tools some train companies put in place, as well as the products and services they propose in order to target and attract responsible travellers.

A review of the most relevant literature on rail tourism will be carried out, staying focussed on the features of the rail-tourism key players and the link between rail tourism and sustainability.

The literature review will be followed by the analysis of the tools and initiatives implemented by three national train companies operating in Europe: the Société Nationale des Chemins de Fer (SNCF), France, the Swiss Federal Railways (SBB-CFF-FFS), Switzerland, and Trenitalia, Italy. The case study analysis will take into account each operator's corporate sustainability strategy, business-to-consumer (B2C) website, and digital or printed promotional material provided to potential or actual passengers.

Eventually, a comparative analysis of the case study will lead to drawing some conclusions on the key elements of the strategies and initiatives applied by train companies to attract responsible tourists and their effective marketing power.

WIT Transactions on Ecology and the Environment, Vol 248, © 2020 WIT Press
www.witpress.com, ISSN 1743-3541 (on-line)
doi:10.2495/ST200101

2 LITERATURE REVIEW

Having briefly mentioned the relation between transport and tourism, the present literature review will focus on the key stakeholders of rail tourism: on the one hand, train companies; on the other hand, train tourists. Particular attention will be paid to a specific group of rail tourists – the responsible travellers – thus paving the way to exploring the link between rail tourism and sustainability.

2.1 Rail transport and tourism

Transport plays a crucial role in tourism development: not only do good transport infrastructures ease the tourists' transfer from their place of origin to their holiday destination, but convenient transport connections and facilities turn out to be a key selection criteria for the choice of the destination itself [1]. Transportation and transport operators are therefore constituent parts of the overall destination product and, as such, they get to influence both the management and the marketing of the destination itself [2].

In the 19th century, the advent of railways made it possible to develop the first popular tourism destinations [3]. Nowadays, high-speed railways are expected to benefit the local economy of the urban areas located all along the railway line by bringing leisure and business tourists and engaging new tourism-driven companies [4]–[7].

2.2 Train operating companies and tourism

Same as for conventional train lines, high-speed railways are exploited by the so-called *train operating companies* (TOCs), i.e. the rail operators responsible for running trains. It is worth mentioning that TOCs do not manage the railway infrastructure facilities: as a matter of fact, the facilities are owned and overseen by one single company – usually a public company, whereas TOCs can be state-owned or fully private – depending on the countries where they operate – and they may or may not compete with other companies offering the same service [8], [9].

While TOCs' core business is to provide train rides, irrespective of the type of travel (business travel, leisure journey, family visits, and thus on), the rail market includes also operators focussed exclusively on rail tourism. Analysing rail tourism in France, Blancheton and Marchi [10] identify three categories of operators associated with rail tourism. The first category makes reference to train companies that own their rolling stock and enjoy dedicated railway lines – including tourist railways, meter-gauge railways and lines with tourism potential. The second category comprises train companies and non-profit associations that possess their rolling stock (e.g. iconic trains like the Orient Express) but make use of railway lines owned and managed by another entity. The third category of rail-tourism operators corresponds to those operators that sell rail-tourism trips and packages, without counting on either their own rolling stock or railway lines.

It is important to highlight that the present paper considers as "train companies" only TOCs. This is because the paper's ultimate objective is to understand through what actions and initiatives conventional train operators – not necessarily used to dealing with tourism – are seeking to attract responsible travellers and meet their needs.

2.3 Rail tourists and responsible travellers

Many different reasons and motivations may lead travellers to move by train.

For some travellers, the train and railways themselves constitute the main attraction, and therefore the main purpose, of their journey [10], [11]. This is the case, for instance, of tourists who choose world-famous train experiences like the Trans-Siberian railway or the scenic railways across the Swiss Alps. Because of their non-mass peculiar tourism preferences, these travellers are representative of *special interest tourism* [12] and become the main market target for the three categories of rail tourism operators identified in France by Blancheton and Marchi [10].

However, even without showing a strong interest in trains and railways, travellers may still prefer the train to other modes of transport. And this for a variety of reasons: the train can be the fastest and most convenient way to reach a destination; the train often takes passengers to the very city centre, thus avoiding traffic congestion and parking issues; the price of a train ticket is potentially more competitive and value-for-money than the price of the alternative flight option, etc. Literature has analysed these elements extensively [3], [4], [5], [7], [13]–[16], also with a view to identifying likely tourism opportunities linked to the flows of travellers – for instance, business travellers – reaching their destinations by train. The same literature points out how some train companies have eventually recognised the tourism potential of train travellers and, based on this, are now competing by complementing conventional train tickets with tourism-related service and products [7], [14], [17].

Regardless of the distance to cover or the price to pay, some could prefer to reach their destination by train simply out of a necessity to slow down and enjoy the journey experience itself (e.g. landscapes, tranquillity, enriching dialogues, etc.). In this sense, travelling by train joins the so-called *slow tourism* [18]–[21]. By opting for a slower means of transport and almost inevitably limiting the number of places and attractions to visit over the trip time, travellers may actually get to live a more intense experience and connect and engage more with the locals. As it was observed by Lyons et al. [22], the vast majority of leisure rail passengers tend to value side and so-to-say intangible experiences such as conversations with travel mates, informal exchanges with co-travellers, window gazing and people watching.

Slow tourism literature shows that the predilection for a slower mode of travel frequently comes alongside environmentally-friendly travel concerns in general. A point of contact that allows to connect slow tourism with responsible travelling and bring this paper into focus.

From a sustainable mobility perspective, the responsible traveller is someone who is willing to replace her/his old mobility habits with more sustainable alternatives, mainly based on the fact that for him/her a more environmentally-friendly lifestyle represents an aspirational goal [23]. The perception of environmental threats is indeed a key factor when it comes to pushing individuals to opt for greener modes of transport [24]. In the current context of growing global concern over the effects of climate change – including the so-called *eco-anxiety* [25], [26] i.e. the spreading anxiety associated with the negative effects of the ecological crisis – and based on the lower carbon footprint of the train as opposed to other modes of transport [27], [28], responsible travellers opting for the rail are likely to grow in numbers. Quite illustrative is the case of Sweden: in parallel with the rise and spread of the *flight shame* movement, the *train bragging* movement is taking hold in the country [26], [29]. As for dissemination and awareness raising on a larger scale, the global press is publishing more and more articles on the possibility to shift from the plane to the rail and the immediate environmental benefits of it, for example [30]–[32].

How responsible travelling is not just about travelling greener, though, is testified to by the feedback given by *flight shame*-movement train travellers in Sweden [29], the aforementioned literature on slow tourism [18]–[21] and studies on green tourists, for example [33], [34]. When responsible travellers engage on a leisure journey, there is evidence that they tend to prefer and enjoy unconventional tourism experiences: they stay in eco-friendly accommodation and seek activities with low impact on the environment and eco-friendly accommodation, they care about local culture and local products and they value the time spent in interacting with the locals. In short, their attitude seems to go beyond the environmental concerns and encompass all facets of sustainability.

Looking at the way traditional train companies are trying to take responsible travellers on board is the purpose of this paper.

3 METHODOLOGY

This paper analyses the strategies and marketing tools, as well as the dedicated products and services, through which traditional train companies are targeting and attracting responsible travellers. The analysis covers three national train companies operating in Europe: the Société Nationale des Chemins de Fer (SNCF) in France, the Swiss Federal Railways (SBB-CFF-FFS) and Trenitalia-Ferrovie dello Stato Italiane (Trenitalia) in Italy. The author singled out these three traditional European train operators based on his familiarity with the rail system in the respective countries and on the languages used in the companies' websites, reports and promotional materials.

The case study analysis takes into account each operator's most recent corporate sustainability strategy, business-to-consumer (B2C) website, and digital or printed promotional material provided to potential or actual passengers. All these sources have undergone consultation and review, in an effort to understand if the three train companies have actually envisaged and implemented specific actions and initiatives for reaching out to responsible travellers.

The findings of the case study analysis will be discussed and some conclusions will be drawn.

4 FINDINGS

The results of the case study analysis have been articulated into three separate sections. The first section will show how each company's corporate sustainability strategy deals with responsible travellers. The second section will illustrate through what channels the three companies are reaching out to responsible travellers. The last section will look at the products and services specifically dedicated to responsible travellers.

4.1 Responsible travellers from a corporate sustainability perspective

It must be said that responsible travellers as such are not mentioned explicitly in any of the three analysed corporate sustainability strategies [35]–[37]. However, the three train companies are clearly committed to both promoting sustainable mobility practices and meeting the expectations of a growing number of environmentally concerned consumers. Another common element of the three corporate sustainability strategies is the reference made to the Paris Agreement on Climate Change and the UN 2030 Agenda for Sustainable Development with its 17 Sustainable Development Goals.

In terms of promoting sustainable mobility practices, the three companies are investing in specific environmental targets such as the gradual reduction of energy, water and waste consumption and the promotion of green procurement and circular-economy solutions. A set of indicators to monitor the achievement of these environmental targets has been put in place. Moreover, the three operators are raising awareness about the train's relative low carbon emissions – in comparison, for example, with the plane. Through some partnership arrangements with external companies, and in order to minimise the journey's overall carbon footprint, the three companies are also giving their customers the option of combining different low-emission modes of transport, e.g. train for long distances and road public transportation for urban or local displacements, directly on their online booking platforms.

In particular, the SBB-CFF-FFS stresses the importance of proposing captivating discounted fares to individual passengers with access to private cars or other more polluting alternatives. Further proof of the commitment to actively promoting sustainability is Trenitalia's yearly campaign *Treno Verde* ("Green Train"), which, in cooperation with the Italian environmental protection association Legambiente, encourages environmentally-friendly practices on board a dedicated train [38].

From a marketing perspective, most of the above-mentioned measures and initiatives are a way to attract new customers and/or reinforce the loyalty of regular or long-term users. No doubt that the target audience comprises environmentally concerned customers, who also constitute the target audience of other initiatives within the companies' marketing plans. As a way of example, the SNCF works to meet the expectations of environmentally conscious customers by serving local organic food on board some of its trains, while Trenitalia encourages slow tourism by revitalising historic railways promoting local culture and local food tradition. Unfortunately, no figures or subscribers' evaluation forms are reported and published in the corporate sustainability strategy of either company, and the same applies to SBB-CFF-FFS.

4.2 Ways to reach out to responsible travellers

The channels through which the three companies seek to reach out to responsible travellers do not essentially differ from the well-established hooks deployed to captivate conventional users, i.e. app, phone number, social media, train station counters, travel agencies, B2C website. All these tools give the opportunity to ask for information and perform bookings. In particular, the B2C website of each of the three companies [39]–[41] is available in several languages and tends to adapt highlights and bargains to the user-selected language. For example, by choosing English as website browsing language, the web user and potential customer is displayed travelling options mainly tailored to non-domestic international travellers. In the case of SNCF and Trenitalia, content customisation applies also to languages other than English: for instance, when consulting the SNCF B2C website in Italian, the train connections between Italy and France come out among the first results. For the time being, what appears to be missing is the customisation of content based on the type of travellers and their main features. In an attempt to become more user-friendly, though, the homepage of the SBB-CFF-FFS' B2B website allows for the selection of the desired travel experience (i.e. city trips, wellness, festivals, etc.), so as to orient the user and adapt the content.

The three train operators reach out to customers also through printed materials: the SBB-CFF-FFS publishes printed brochures and information leaflets, as well as an on-board magazines; the SNCF seems to invest more in information leaflets, whereas Trenitalia mainly counts on on-board magazines available on long-distance high-speed trains. According to the promotional printed material collected by the author, most information on sustainable mobility and sustainable travelling can be found in the on-board magazines: from enjoyable articles suggesting natural immersion gateways to visual adverts for cultural discovery weekends.

Speaking of promotion of sustainable mobility practices, it is worth reporting Trenitalia's commitment to informing their high-speed passengers of the estimated amount of kilograms of carbon dioxide saved thanks to her/his choice to take the train instead of the car or the plane. The amount of saved CO_2 is revealed on both Trenitalia's digital and printed version of the train ticket.

4.3 Products and services tailored to responsible travellers

On top of conventional train tickets, the three train companies taken into consideration in this study are giving then train traveller the opportunity to build up a full – and green – tourism experience by proposing additional multi-operator and multi-experience travel combinations. Multi-operator and multi-experience combinations which are often the result of some partnership agreements with external providers and comprise hotel booking, car rentals, luggage shipping or even car-/bike-/scooter-sharing in a given geographical area. SBB-CFF-FFS and Trenitalia go even further, by proposing the direct booking of museum tickets and guided tours.

The provision of environmentally- and user-friendly options, e.g. combination of train and other low-emission travel or cultural experiences, show how the three train companies strive to open up rail tourism in general and responsible rail travellers more in particular. The three companies' awareness of the preferences of responsible travellers can be inferred from the analysis of their corporate sustainability strategy, with relevance given to aspects like the provision of locally-sourced organic food on board (SNCF) and the discovery of regional culture and traditions through the revitalisation of historic railways (Trenitalia). In the case of SBB-CFF-FFS, the care for sustainable travelling is reflected in both the B2C website and the printed materials, where the Swiss operator adverts day tours or short holidays by rail evoking nature, eco-friendly activities and responsible tourism in general.

5 DISCUSSION

The analysis of the sustainability strategy, the B2C website and the promotional material of the three train companies considered in the context of this paper shows that train operators have become aware of the growing concern over climate change and environmental protection. It can be said that train companies are committed to actively contributing to the implementation of the 17 Sustainable Development Goals and the Paris Agreement on Climate Change. In line with this, they have set targets for on-board energy, water and waste reduction and they have developed specific indicators to monitor and assess the achievement of these targets. Awareness-raising initiatives – such as the promotion of sustainable mobility practices through their on-board magazines – have been also elaborated and implemented.

In addition to raising awareness about environmental issues, these steps towards sustainability can also be looked at as powerful marketing tools to attract new customers and/or increase the loyalty of existing ones. In times of growing global concern over the effects of climate change and *eco-anxiety*, rail operators seem to be willing to capitalise on the environmentally-friendly reputation of modern trains. To emphasise the low carbon footprint of train travelling as opposed to cars or planes is an influential way to reach out to responsible travellers, i.e. travellers not only fighting for green mobility but also looking for slower, sustainable and locally-embedded tourism opportunities.

The channels used by train companies to get in contact and interact with responsible travellers are the same as the ones used to dialogue with conventional customers, i.e. apps, phone numbers, social medias, train station counters, travel agencies, the B2C website and digital or printed promotional materials. Although not specifically designed for responsible travellers, these channels display very interesting products and services for tourists planning sustainable trips or even vacations. This can be the case of cultural discovery train journeys offers and train- and bike-sharing combinations. In spite of all this, though, it is difficult – if not impossible – at this stage to gather figures on how many travellers are buying or considering buying these products and services.

The information currently available in the sources consulted by the author in the context of this research does not allow for a comprehensive analysis of the size and evolution of the responsible rail tourism market segment. Corporate reports limit themselves to presenting figures related to train rides and train passengers in general, without indicating the reasons, motivations or purposes behind individual or group trips. This might depend on the lack of surveys and questionnaires aimed at collecting and clustering users' features. Moreover, the companies' B2C websites do not provide for user-dedicated access or content, thus limiting customised content to language-driven results only.

6 CONCLUSION

The growing global concern over the effects of climate change is seriously affecting people's preferences in terms of mobility. Despite the step back in the use of public transport in the short term mainly due to COVID-19 [42], sustainable modes of transport are expected to come out on top in the long term. In addition to opting for more environmentally-friendly means of transport, though, responsible travellers are also looking for slower, locally-embedded and overall sustainable tourism experiences.

This work has shown that traditional train companies are aware of the considerable potential of the train in attracting responsible travellers. They have set environmental targets and they are actively raising awareness about the lower carbon footprint of train trips as opposed to other mobility options. Furthermore, on their conventional ticket booking platforms they are offering a number of complementary products and services, e.g. train and bike combinations and historic railway journeys, likely to catch the eye of both ordinary tourists and responsible travellers.

Yet, conventional train companies like the ones examined within the framework of this paper do not seem to be fully equipped to identify the needs of responsible travellers and monitor the evolution of their market segment. The figures published by train operators report on train rides and train passengers only, without specifying the train users' characteristics and reasons for travelling. To the same extent, no online survey or any similar tool have helped to profile the visitors and users of the companies' B2C website. Such an initiative could certainly lead to more targeted marketing decisions.

Train operators would certainly benefit from further research on the topic, too. There is a need for analysis of more case studies on rail tourism, if possible including interviews with both train operators' managers and users. Interviews with train companies' managers in particular should aim to deepen the knowledge about (i) the current and future strategies targeting responsible travellers, (ii) the potential networking and partnership activities with specialised tour operators, and (iii) the measures and indicators used to monitor the performance of responsible travellers-related initiatives.

REFERENCES

[1] Prideaux, B., The role of the transport system in destination development. *Tourism Management*, **21**(1), pp. 53–63, 2000.

[2] Buhalis, D., Marketing the competitive destination of the future. *Tourism Management*, **21**(1), pp. 97–116, 2000.

[3] Michniak, D., Role of railway transport in tourism: Selected problems and examples in Slovakia. *Quaestiones Geographicae*, **35**(4), pp. 107–120, 2016.

[4] Albalate, D. & Fageda, X., High speed rail and tourism: Empirical evidence from Spain. *Transportation Research Part A: Policy and Practice*, **85**, pp. 174–185, 2016.

[5] Delaplace, M., Pagliara, F., Perrin, J. & Mermet, S., Can high speed rail foster the choice of destination for tourism purpose. *Procedia-Social and Behavioral Sciences*, **111**(1), pp. 166–175, 2014.
[6] Masson, S. & Petiot, R., Can the high speed rail reinforce tourism attractiveness? The case of the high speed rail between Perpignan (France) and Barcelona (Spain). *Technovation*, **29**(9), pp. 611–617, 2009.
[7] Yang, Y., Li, D. & Li, X., Public transport connectivity and intercity tourist flows. *Journal of Travel Research*, **58**(1), pp. 25–41, 2019.
[8] Jeffcott, S., Pidgeon, N., Weyman, A. & Walls, J., Risk, trust, and safety culture in UK train operating companies. *Risk Analysis*, **26**(5), pp. 1105–1121, 2006.
[9] Luan, X., Corman, F. & Meng, L., Non-discriminatory train dispatching in a rail transport market with multiple competing and collaborative train operating companies. *Transportation Research Part C: Emerging Technologies*, **80**, pp. 148–174, 2017.
[10] Blancheton, B. & Marchi, J.-J., The three systems of rail tourism: French case. *Tourism Management Perspectives*, **5**, pp. 31–40, 2013.
[11] Nering, N. & Feger, J.E., Competitiveness factors of tourist train operator in Brazil. *Podium Sport, Leisure and Tourism Review*, **8**(1), pp. 81–97, 2019.
[12] Trauer, B., Conceptualizing special interest tourism: Frameworks for analysis. *Tourism Management*, **27**(2), pp. 183–200, 2006.
[13] Dallen, J., The challenges of diverse visitor perceptions: rail policy and sustainable transport at the resort destination. *Journal of Transport Geography*, **15**(2), pp. 104–115, 2007.
[14] Delaplace, M. & Dobruszkes, F., From low-cost airlines to low-cost high-speed rail? The French case. *Transport Policy*, **38**, pp. 73–85, 2015.
[15] Moyano, A., Rivas, A. & Coronado, J.M., Business and tourism high-speed rail same-day trips: Factors influencing the efficiency of high-speed rail links for Spanish cities. *European Planning Studies*, **27**(3), pp. 533–554, 2019.
[16] Wu, N., Qiu, Y. & Chen, L., Research on marketing strategy of intercity railway in Wuhan metropolitan area. *Advances in Economics, Business and Management Research*, **68**, pp. 650–657, 2019.
[17] Prideaux, B., Tracks to tourism: Queensland rail joins the tourist industry. *International Journal of Tourism Research*, **1**(2), pp. 73–86, 1999.
[18] Caffyn, A., Advocating and implementing slow tourism. *Tourism Recreation Research*, **37**(1), pp. 77–80, 2012.
[19] Dickinson, J.E., Robbins, D. & Lumsdon, L., Holiday travel discourses and climate change. *Journal of Transport Geography*, **18**(3), pp. 482–489, 2010.
[20] Lumsdon, L.M. & McGrath, P., Developing a conceptual framework for slow travel: A grounded theory approach. *Journal of Sustainable Tourism*, **19**(3), pp. 265–279, 2011.
[21] Oh, H., Assaf, A.G. & Baloglu, S., Motivations and goals of slow tourism. *Journal of Travel Research*, **55**(2), pp. 205–219, 2016.
[22] Lyons, G., Jain, J. & Holley, D., The use of travel time by rail passengers in Great Britain. *Transportation Research Part A: Policy and Practice*, **41**(1), pp. 107–120, 2007.
[23] Holden, E., Banister, D., Gössling, S., Gilpin, G. & Linnerud, K., Grand narratives for sustainable mobility: A conceptual review. *Energy Research and Social Science*, **65**, 2020.

[24] Wang, S., Wang, J. & Yang, F., From willingness to action: Do push-pull-mooring factors matter for shifting to green transportation? *Transportation Research Part D: Transport and Environment*, **79**, 2020.

[25] Ingle, H.E. & Mikulewicz, M., Mental health and climate change: Tackling invisible injustice. *The Lancet Planetary Health*, **4**(4), e128–e130, 2020.

[26] Mkono, M., Eco-anxiety and the flight shaming movement: Implications for tourism. *Journal of Tourism Futures*, 2020.

[27] Givoni, M., Environmental benefits from mode substitution: Comparison of the environmental impact from aircraft and high-speed train operations. *International Journal of Sustainable Transportation*, **1**(4), pp. 209–230, 2007.

[28] Wickizer, B.J. & Snow, A., Rediscovering the transportation frontier: Improving sustainability in the United States through passenger rail. *Sustainable Development Law and Policy*, **11**(1), pp. 12–16, 2011.

[29] Jacobson, L., Åkerman, J., Giusti, M. & Bhowmik, A.K., Tipping to staying on the ground: Internalized Knowledge of climate change crucial for transformed air travel behavior. *Sustainability*, **12**(5), pp. 1–18, 2020.

[30] Birnbaum, M., Europe's flight-shame movement has travelers taking trains to save the planet. *The Washington Post*, 2 Aug. 2019. www.washingtonpost.com/world/europe/europes-flight-shame-movement-has-travelers-taking-trains-to-save-the-planet/2019/08/02/1bd38486-ac96-11e9-9411-a608f9d0c2d3_story.html. Accessed on: 2 Jun. 2020.

[31] Gardner, N., No need to fly: Three European trips easier by train. *The Guardian*, 13 Jan. 2020. www.theguardian.com/travel/2020/jan/13/rail-itineraries-europe-tarian-holidays-austria-netherlands-coast-france-burgundy-vineyards. Accessed on: 2 Jun. 2020.

[32] Vaudano, M. & Sénécat, A., Quels trajets intérieurs en avion pourraient être interdits en cas d'alternative en train? *Le Monde*, 8 Jun. 2019. www.lemonde.fr/les-decodeurs/article/2019/06/07/la-majorite-des-vols-interieurs-pourrait-se-faire-en-train-et-ce-serait-bon-pour-l-environnement_5472992_4355770.html. Accessed on: 2 Jun. 2020.

[33] Dolnicar, S., Crouch, G.I. & Long, P., Environment-friendly tourists: what do we really know about them? *Journal of Sustainable Tourism*, **16**(2), pp. 197–210, 2008.

[34] Falk, M. & Hagsten, E., Ways of the green tourist in Europe. *Journal of Cleaner Production*, **225**, pp. 1033–1043, 2019.

[35] SBB-CFF-FFS SA, Rapporto di gestione delle FFS 2019. https://company.sbb.ch/content/dam/internet/corporate/it/medien/publikationen/geschaeftsbericht/SBB-Geschaeftsbericht-2019.pdf.sbbdownload.pdf. Accessed on: 6 Jun. 2020.

[36] SNCF, 2018 Corporate social engagement report, Direction du développement durable. https://medias.sncf.com/sncfcom/finances/Publications_Groupe/SNCF-RSE_english.pdf. Accessed on: 25 May 2020.

[37] Ferrovie dello Stato Italiane, Group Strategy, Planning, Innovation and Sustainability Department, 2018 Sustainability report: The journey goes on. www.fsitaliane.it/content/dam/fsitaliane/en/Documents/commitment/sustainability-report/2018-sustainability-report-/2018SR%20-%20Executive%20Summary.pdf. Accessed on: 5 Jun. 2020.

[38] Treno Verde, Legambiente Onlus. www.trenoverde.it. Accessed on: 8 Jun. 2020.

[39] OUI.SNCF, Société Nationale des Chemins de Fer (SNCF). https://en.oui.sncf/en. Accessed on: 8 Jun. 2020.

[40] SBB-CFF-FFS, Swiss Federal Railways (SBB-CFF-FFS). www.sbb.ch/en/home.html. Accessed on: 8 Jun. 2020.

[41] Trenitalia, Trenitalia-Ferrovie dello Stato Italiane. www.trenitalia.com. Accessed on: 8 Jun. 2020.

[42] Leenen, M., Remainder of 2020 looks bleak for rail. *International Railway Journal*, 3 Jun. 2020. www.railjournal.com/in_depth/remainder-of-2020-looks-bleak-for-rail. Accessed on: 11 Jun. 2020.

UNDERSTANDING THE SUSTAINABILITY STANCE OF MICRO AND SMALL-SIZED ACCOMMODATION OWNER-MANAGERS TO ENTER INTO (SUSTAINABLE) ENTREPRENEURSHIP

FEMKE A. VRENEGOOR[1,2], GJALT DE JONG[2] & ELENA CAVAGNARO[1]
[1]Academy of International Hospitality Research, NHL Stenden University, The Netherlands
[2]Centre for Sustainable Entrepreneurship University of Groningen/Campus Fryslân, The Netherlands

ABSTRACT

This research looks at the motivates and values of owner-managers of micro and small-sized accommodations for entering into the accommodation business and implementing CSR measures in it. According to theory, it may either be because of self-enhancement motives, for self-transcendent motives, or a combination of the two. These different motives may lead to the entrepreneur implementing sustainability measures in his operations, only the extent and depth to which this is done depends on the underlying values of the person. Explorative semi-structured interviews with the owner-managers of micro and small-sized accommodations were carried out. Results show that indeed the primary reason for entrepreneurs to enter into business is for self-enhancement reasons. Notwithstanding, a number of entrepreneurs mention both self-enhancement and self-transcendence motives. Most entrepreneurs do not mention sustainability as part of their business goal, but have implemented several CSR measures. This goes from basic actions like LED lights and separating waste, until solar panels and waste prevention. When referring to sustainability, the entrepreneurs mostly explain it as actions that are good for the environment. Actions that are good for society are not seen as belonging to sustainability, even though the entrepreneur undertakes several actions that add value to it. Further research is needed to get a better understanding of the types of CSR measures that are implemented in their accommodations, and how this connects to the dominant values of the individual.

Keywords: tourism entrepreneurs, accommodation owner-managers, sustainability stance, motives, values, CSR measures, SME's.

1 INTRODUCTION

The aim of the research presented in this paper is to develop a deeper understanding of the motives of micro and small-sized accommodation owner-managers to enter into business, their sustainability stance and their dominant values. Value theory is used to explain whether patterns in behaviour can be identified. This research thus contributes to the limited research on the intersection between tourism SME's, sustainability and values. This is important, as the tourism industry is seen as having a responsibility to behave morally and ethically [1] and values are ultimate antecedents of pro-environmental behaviour [2]. Values have been shown to decide the business operations for some owner-managers in tourism [3], though little is known about their motives to behave sustainable and the CSR measures they implement [4], especially for SME's [5].

For future generations to be able to have the same quality of life as we currently have in the Western world, it is necessary to take global environmental and social action. The United Nations (UN) created a set of 17 Sustainable Development Goals (SDGs) to end poverty, protect the planet and ensure prosperity for all [6]. A clear role for companies is shown in the SDGs, asking for production of goods and services in a responsible manner, thus "doing more and better with less" [6]. This is necessary, as companies are seen by many to be the most important protagonists of the current unsustainable state of society [7], [8].

Globally, the tourism industry is an important contributor to the economy and has seen steady growth over the last few years. In 2017, tourism growth was higher than that of the global economy [9]. In the European Union, it contributes 10% to its GDP and provides direct and indirect jobs for 26 million persons [10]. In the Netherlands, the direct contribution of tourism to the GDP was 4,4% in 2018, and 6,3% of all jobs were in tourism [11]. According to UNWTO [10], 33% of all tourism establishments are hotels and similar accommodations and in the Netherlands 95% of accommodations fall in the micro (less than 10 rooms) and small (less than 50 rooms) size category [12].

Small firms play a key role in the sector [5], [10], but are seen to be lagging behind in their implementation to reduce their negative impact to climate change as to reach the SDGs. As Thomas et al. put forward, SMEs in tourism "have largely failed to embrace the growing environmental agendas of the public sector" [5, p. 969]. Current tourism development is unsustainable [13], and is responsible for approximately 5% of global CO_2 exhaust [13]. Within the tourism industry, accommodations are responsible for 21% of the impact on climate change [13], and in the Netherlands they are even responsible for about 30% of the CO_2 exhaustion of tourism [14].

To be sustainable, an organization should "manage its resources in such a way that economic, social and environmental benefits are maximized in order to meet the need of the present while protecting and enhancing opportunities for future generations" [15, p. 4]. One way for an organization to take care of this, is by implementing Corporate Social Responsibility (CSR) principles in their company. CSR is defined as the active and voluntary contribution of an enterprise to environmental, social and economic improvement [16], and is concerned with a company's obligations to be accountable to all its stakeholders when operating and undertaking business activities [15]. When speaking about CSR in this paper, all the company's sustainability efforts are meant. The focus of CSR research has been particularly on larger companies. SMEs have received little attention [4], [17]. It is the owner-manager in such accommodations that shapes the organisations strategy [18], and existing research has "failed to understand the reasons behind the possible cognitive constraints that can affect the sustainable orientations of decision-making by SME hotel entrepreneurs" [19, p. 146].

There is an ongoing debate about whether the owner-managers of tourism SME's may be called entrepreneurs or not. It is not a goal of this paper to go into this matter. The term entrepreneurs will be used for the owner-managers, as according to Drucker [20] entrepreneurship is a form of behaviour. In academic literature some attention has been given to the antecedents of this behaviour, i.e. individual hotel entrepreneurship. Especially the motives, attitudes and business challenges of such entrepreneurs have been investigated [17]. Though attitudes are one of the factors that decide behaviour [2], literature shows that there is a gap between attitude and behaviour [2]. Values are considered to be stable and ultimate antecedents of behaviour, and might therefore predict not contemporaneous and future choices [2]. They are defined as "desirable trans-situational goals varying in importance, which serve as guiding principles in the life of a person or society" [21, p. 21]. Values guide pro-environmental behaviour such as energy reduction [2], and help to understand why a person will take certain sustainable actions, but will not take others. Limited attention has been given to values as an antecedent of accommodation entrepreneurship, though they are the ultimate antecedent of behaviour, whereas motivations are immediate antecedents [2].

Hence, it makes a lot of sense to study motives and values together. For this research, particularly the values the individual holds as well as their motives to operate a hotel are interesting, because the dominant values, type of motives and the strength they have determine the goals the individual has for their business [17]. Operating in a sustainable manner may be one such a business goal. Therefore, we present the findings of our research amongst the owner-managers of micro and small-sized accommodations in the Netherlands. We asked them about their motives to become an entrepreneur, their value orientations and the CSR measures implemented by them and hope to thus help close these knowledge gaps.

2 THEORETICAL UNDERPINNING

Thanks to different streams of entrepreneurship literature, we know why some individuals implement CSR in their companies, and others do not. This section will first define traditional entrepreneurship before going into lifestyle entrepreneurship. After, the role of values will be explained in relation to motives to become an entrepreneur to understand from which value orientation that behaviour stems.

Entrepreneurship, in the Schumpeterian sense of the concept, refers to "creative destruction", as it seeks to replace existing products or services in the market place with a superior one [22]. The ability to innovate, to create a new product or service and introduce it successfully to the market place to grow one's business, is seen as a distinguishing feature for entrepreneurs [8]. Opportunity recognition, innovation and risk taking are seen as essential [7], [8]. The afore mentioned definition finds its basis in economic studies, and following it, the owner-managers of micro and small-sized tourism firms would not be considered entrepreneurs due to their lack of having economic goals. Yet, tourism research repeatedly speaks about tourism entrepreneurs, regardless of the size of the organization. This may be explained by considering the specific context of this sector. As Skokic and Morrison put it "this [economic] perspective has failed to explain the motivation of numerous small businesses in tourism which defy models of economic rationality, that of lifestyle entrepreneurs" [23, p. 157]. Lifestyle entrepreneurs are identified as those entrepreneurs who own and operate a business closely aligned with their personal values, beliefs, interests and passions [8], [30]. For them, success of their organization is not expressed in economic terms, but rather in their continued ability to maintain their chosen lifestyle [24]. This puts them at the other end of the renowned continuum across from for-profit/commercial entrepreneurs. However, lifestyle entrepreneurship is still a rather fuzzy concept, and there is some indication that there are more nuances in the motives for entrepreneurship [3], [5], [25].

Besides lifestyle entrepreneurs, another relatively new type of entrepreneur has been identified in literature, namely the sustainable entrepreneur. Such individuals use the opportunity arising from a certain market failure (in this case the unsustainability of the world) as a specific organisational goal to reach value creation on the triple bottom line of people, planet and profit [7], [8]. Here there is an explicit addition of creating social and environmental value, thus this type of entrepreneurs is motivated by self-enhancement values and self-transcendent values. One such an example in the hospitality industry is the Feldmilla hotel in Italy. This is a small family run enterprise which is fully climate neutral and their business goals are connected to society and the environment. However, it remains unclear if all lifestyle entrepreneurs also pursue sustainability goals, such as the Feldmilla hotel.

What is clear, is that next to profit, socioeconomic and environmental values play a role in the decision-making process of at least lifestyle entrepreneurs in tourism [3], and that "applying psychological and sociological lenses" has value when trying to understand the behaviour of individuals commercially letting out accommodation [5, p. 966]. Font et al. [4] offer insight into the implementation of CSR measures by SME tourism business by identifying three frames for engaging in them; cost-reduction competitiveness (taking measures that lead to operational and internal benefit), societal legitimization (taking measures that are visible or expected by others) and lifestyle-value drivers (taking measures due to personal choices and out of habit). It may even be argued that operating one's organisation in a sustainable manner may be an outcome of effectuation, rather than a purposeful goal [26]. However, the causation and effectuation theory seems to be largely ignored in tourism and lifestyle research.

It is evident that not all entrepreneurs behave equally sustainable, and that they have different motives to engage in the implementation of CSR measures, if they do so at all. It is necessary to investigate this further due to the considerable CO_2 exhaust caused by accommodations, and the large percentage of SME's in the sector. If all these accommodations were to become more sustainable, this would be a significant improvement of the sustainability stance of the tourism industry.

Environmental psychology, social-psychological theory and cognitive theory offer answers to understanding these different sustainability stances by looking at pro-environmental behaviour, and the underlying values guiding such behaviour. Many definitions of pro-environmental behaviour exist, but in this research, it is understood as behaviour that adds value to humans and the environment. Humans are part of the environment and that acting pro-environmentally therefore equals actin pro-socially. Values are a person's guiding principles to reach a desirable trans-situational goal in life [21]. Since they guide life goals, they guide attitude and behaviour to reach that life goal. Therefore, they are important to consider when discussing pro-environmental behaviour. Schwartz' theory reflects that every person holds a number of values at the same time, and the relative importance give to those values differ per person and per circumstance. Values are not about specific action, but rather exceed them, and thus serve as an individual's standards based on which (s)he assesses what is right or wrong, what should be done or not done, and so on. Since each person holds different values, that may be in conflict with each other, a trade off occurs between competing values. Similar values are referred to as value types, and clusters of compatible value types are called value orientations [2].

Values relevant to pro-environmental behaviour can be split in two categories, namely self-enhancement values and self-transcendence values. The first category concerns the overarching value orientation as valuing power, achievement and hedonism, and the second category concern the overarching value orientation for universalism and benevolence [21]. Theoretically, a person with dominant self-enhancement values may perform as pro-environmentally, as persons with dominant altruistic and biospheric values, if those actions align with their self-interest. However, individuals with strong self-transcendence (i.e. altruistic and biospheric) values are more likely to show more salient pro-environmental behaviour [2]. Notwithstanding, it has been shown that general beliefs in support of sustainability, intentions and behaviour are positively connected to biospheric values, and that they are negatively connected to egoistic values [2].

The entrepreneurial behaviour to grow one's business can be seen as behaviour according to self-enhancement values [21], connected to an egoistic value orientation [2]. Becoming an entrepreneur because of the freedom it allows oneself, and because of the status it gives fall under the above explanation. Becoming an entrepreneur to be able to provide for your family, or to help better the economy in one's area are at least partially motivated by altruistic (thus self-transcendence) values. Research using cognitive theory to understand the motives for entrepreneurs in the accommodation industry to enter into business describe various reasons. Some of the most important ones are financial motives, personal satisfaction, independence/working for oneself, and fulfilling a desire to open a hotel [17], [18]. All of these motives fit the definition of self-enhancement values, despite the often-conscious rejection of economic and business growth opportunities in favour of maintaining a certain lifestyle [3].

Entrepreneurs with high hedonic and gain values thus have a lower propensity to implement sustainability measures that require effort. This may form a barrier to more stringent implementation of sustainability in accommodations. This barrier is aggravated further by the nature of hospitality: a hotelier's role is to provide the luxury, comfort and service quality that a guest is seeking. These entrepreneurs do not want to hinder guests in their organisations with striving for more sustainable operations [33], even though guests themselves appreciate the green efforts of a hotel [28]. One could say that the aforementioned reflects a cognitive dissonance in accommodation entrepreneurs, where on the one hand they wish to add value to sustainability via their operations, but on the other hand do not want to decrease the guests' experience during their stay.

Values are ultimate antecedent of behaviour, motivations immediate antecedents. A better understanding of the motivation of an entrepreneur to operate an accommodation, the CSR measures they implement and their dominant values will allow to better comprehend their sustainability stance. For this reason, the next section explains how this data was collected.

3 METHOD

To collect data, explorative face to face semi-structured interviews were completed. This qualitative approach was chosen because it allows to openly examine the experiences of entrepreneurs, the meaning they give to their business and the choices they make. To avoid socially desirable answers, the respondents were not informed about the overall goal of this research as not to frame them towards sustainability. Rather, the hotelier was asked for their motives to become an entrepreneur in the accommodation industry. This allowed to see if the respondent comes up with sustainability as a motive for operating their business themselves, or only after explicitly being asked for it. This is necessary, as it gives an indication of the prioritized values of the hotelier. To support this goal, a value card game with the 16 values significant for pro-environmental behaviour, as determined by Schwartz [21] and Steg et al. [27] was used. One additional universal value was added to this card game, as entrepreneurship theory shows it is of major importance, namely freedom. In total, the card game contained 17 value cards. During the interview, the owner-manager was asked to select the 5 values that (s)he finds most important, and to rank them hierarchal from most important to least important. A picture of this ranking was taken for analysis purposed, and the five selected cards were used for further discussion with the following interview questions.

A combination of convenience and purposive sampling was used to collect the data. This was done as entrepreneurs in any case are a hard-to-reach group, and even more so in the accommodation industry. Using these types of sampling helped getting access to the needed population for an interview. Owner-managers of accommodations with fewer than 50 rooms, which in the Netherlands' make up more than 90% of the market [12], either independent or franchise, were selected. Interviews continued until saturation was reached. In total, 17 interviews were collected. Of those, just over half were with women. Most of the respondents were middle aged. Typically, the accommodations were Bed & Breakfasts or small hotels with a restaurant. In this sample, the number of rooms to let ranged from 2 to 41 rooms, showing a spread amongst the size categories.

The interviews were transcribed true verbatim with the help of research assistants under the supervision of the main author. Standard confidentiality practices were adhered to. The data was analysed using a deductive approach. To avoid interviewer bias, coding and analysis of the transcripts was done by two researchers separate from each other. This process of triangulation increases the validity of the research outcomes by avoiding conceptual interpretation. Using an iterative process following DeCuir-Gunby et al. [29], a codebook was created based on data-driven open and axial coding of the interview transcripts.

Intercoder reliability was tested by two researchers using the agreed upon codebook to code two randomly selected interview, and compare results. The Kappa coefficient showed almost perfect agreement.

4 FINDINGS

Below the findings from the data are presented following the three main themes deriving from the interviews, namely motives, sustainability and values.

4.1 Motives

Entrepreneurs enter into business due to different motives. A range of motives was found in the data, but in general two categories can be identified. The first contains those entrepreneurs with a passion for the accommodation industry, seeking a professional career in it. The second category contains entrepreneurs who have a passion for something else. This "something else" leads to a plethora of reasons to enter into business. Here one finds the lifestyle entrepreneurs that seek to balance their personal interests, needs of their family and their business career [18], [30], [31], the opportunity-driven entrepreneurs that faced some business opportunity and choose to become an accommodation entrepreneur because of it [32] and those for whom the accommodation is a means to fulfilling another main business goal. Though there might be an overlap between these reasons for going into the accommodation business, the entrepreneurs belonging to this category did not have owning-managing an accommodation as their primary goal. Therefore, this author calls entrepreneurs in this category "accidental hoteliers", which doesn't mean the accommodation form cannot be a B&B.

In the category with accommodation entrepreneurs with a passion for the industry mostly owner-managers of hotels with a restaurant are found (I9, I10, I14 and I17), though there is one hotel without a restaurant (I15) and one B&B (I6). When asked for their motives to enter into business, several interviewees mentioned that their parents owned a hotel, and they therefore grew up being and working in a hotel (I10, I14 and I17). One was even the 5th generation of hotel owners in his family (I17). Working in a hotel was therefore "taught from the nursery" (I10). Interestingly, over half of these entrepreneurs did venture into other lines of work before committing to a career as the owner-manager of an accommodation (I9, I14 and I17). As one person says "the hospitality sort of is an addiction" and that it had been "a little girl's dream" to own a hotel (I9). Others (I10 and I15) never left the hotel industry and went straight from a hospitality education into working in their hotel (I10), or had had a long career as general manager in hotels and were looking for a new challenge which they found in owning their own hotel (I15). Smaller types of accommodation owners may also fit in this category. For example, when moving to another part of the country, one B&B owner-manager purposely looked for a house that gave them the possibility to operate a B&B from it, with enough space to share with guest. Before actually purchasing the property, they looked into the possibility of getting the required licenses (I6). Similar findings were reported by Lashley and Rowson who found that accommodation owner-managers reported having a life-long dream of owning a hotel, or thought that they would enjoy operating one [18].

In the category with a passion for something else, the accidental hoteliers, most entrepreneurs are owner-mangers of an accommodation in support of another primary goal. Frequently, this other goal is having a restaurant (I7, I8, I13 and I16). As one interviewee stated "I had the ambition to own a restaurant, and coincidentally this one came with a hotel" (I7). Others said they were looking for a new challenge after owning a previous restaurant (I13). Though the accommodation was not their primary goal, these entrepreneurs did see the merit of it being connected to their restaurant. The guests that stay in the hotel, also eat in the restaurant (I7, I16). One entrepreneur referred to it helping him get through the low season of the restaurant, as it allowed him to sell package deals (I16).

For others, the main motivation is not found within the hospitality or related business. Amongst the accommodation owner-managers livestock and poultry farmers are found (I2 and I5). The poultry farmer wanted to build new stables, and bought land from her neighbours as to be able to do so (I2). Since the land purchased had a "perfectly good house" on it, they decided to "do something with it, as leaving it empty would be a waste" (I2). For the livestock farmer operating the B&B was the result of his search for a way to supplement the income from his farm, thus keeping this primary business viable for the future (I5).

One interviewee said that he planned on becoming an entrepreneur, as he didn't want to work for a boss for long. However, "it didn't have to be in the hospitality industry" (I12) and he even didn't want to remain working in it for too long. However, when the previous owner of his hotel quit, this opportunity was offered to him, and he thought it would be nice. Another emerging motive for becoming an accommodation owner-manager is maybe the purest form of "accidental hotelier". These individuals didn't plan to go into the accommodation industry or even the hospitality industry. They simply had space available in their homes (I1, I3, I4). It came to be organically, out of the desire to remain living in their house after a divorce, or avoiding having a gap in their resume while being a stay-at-home-mom.

Lifestyle reasons for becoming an entrepreneur are found in both the passion for the accommodation industry, and the passion for something else category. Some female owner-managers mentioned that becoming an accommodation entrepreneur allowed them to work from home, which combined well with a family with (young) children (I1, I4 and I9). Spending quality time with family and friends is recognized as one of the lifestyle motives [30], [32]. Another commonality is not necessarily looking for profit maximization, but rather for alignment between one's personal interests and their business career [19], [30]. Few of these accommodation owner-managers were actively looking for expansion possibilities. As one entrepreneur states, "right now it is enjoyable and it is not an obligation" (I2). Expanding would mean having to hire personnel, whereas now she could manage on her own. Not aiming to maximize economic gain is recognized in tourism and hospitality literature as fitting with lifestyle reasons for operating a small business [3], [30]. On the other hand, some owner-managers did want to expand and are on the path of doing so. As one would expect based on literature [4], none of the entrepreneurs mentioned sustainability as the main reason for becoming an entrepreneur.

4.2 Sustainability

When asked to explain what sustainability means, most respondents state that it is about caring for the environment and not wasting its resources (I2, I3, I5, I8, I9, I10, I17), or preventing global warming (I10). Some individuals mention it is about durability, considering what it better for the environment for the future (I8, I9, I13). Several owner-managers sum up measures that are good for the environment and why they are good for the environment, rather than explaining what sustainability entails (I1, I3, I4, I6), where some referred to considering the impact of ones actions (I1, I5, I14) and "minimizing your footprint" (I14). Interestingly, only one entrepreneur specifically mentions social sustainability when discussing sustainability in his company (I8) where in his case it concerned the wellbeing of his employees. Another entrepreneur refers to the importance of sustainable relationships "with employees, guests and suppliers" (I17). Several entrepreneurs undertake measures that add value to the people aspect of sustainability, but do not recognize it as such. A typical example is sourcing products locally which they see as better for the environment due to less transport, but do not realize the social side of it. This situation found by the author is not exceptional, as in hospitality, but also in general, most initiatives are geared towards the environmental side of sustainability [33]. As pointed out by Elkington and Rowlands especially the value addition to the social dimension has been overlooked [34], and still many businesses focus on the environmental and economic dimensions only [33].

Though not necessarily indicative of a negative propensity towards sustainability, some entrepreneurs stated that they dislike the word sustainability (I2, I8, I11, I17). These interviewees feel the word is used as an umbrella term for many different actions, is "used too much" (I8) and that it is a "relative concept" (I11) used with a broad meaning (I17). Alcott states that sustainability is "one of the least meaningful and most overused words in the English language" [35, p. 246], and the author of this paper is confident that the same goes for the usage of the Dutch version of the word in the Netherlands.

The nature of providing hospitality is strongly connected to the pampering and comfort of the guest. When sustainability reduces such comfort, it is seen as conflicting with the nature of the entrepreneurs' business (I1, I7, I8, I9, I11). This puts entrepreneurs in a position where they have to consider which they will prioritize; being environmentally friendly or providing the guests with the pampering they expect [36]. Common issues described are having to maintain a comfortable temperature in the accommodation and the pertaining restaurant (I7, I9), as well as daily clean towels and bed sheets (I7, I9). Most of the entrepreneurs prioritized the comfort of the guest above reducing the use of natural resources, though one owner-manager stated that she refused to tumble dry the towels, as she though it to be unsustainable (I1).

Entrepreneurs state that when faced with a decision, they do consider if it is possible to choose a (more) sustainable option. They evaluate the cost and benefit of doing so, and then make an informed decision. This goes for basic measures such as replacing the LED lights, but also for more complex ones such as insolation. Typically, the occasion for evaluating the more complex measures is at the time of a planned refurbishment or new construction (I1, I3, I8, I9, I10, I14). One interviewee explains that they are currently renovating their hotel rooms, and are looking at if it can be done sustainable, and how (I14). Another mentions that they will renovate their kitchen, and in anticipation future law prohibiting the use of gas, will choose another form of cooking (I8). This coincides with literature, where this same moment was indicated as an opportune one for installing fixtures that helped the hotel reduce their utility usage [37]. Besides the added value to the environment when the business is updated according to sustainability standards, entrepreneurs state it is also about "keeping with the times", as well as keeping your company viable for potential future sale (I2, I17) At least the latter is an economic motive as well.

In general, cost is mentioned as typical barrier for not choosing the sustainable option, as well as it costing too much time and effort to implement (I1, I2, I10, I12, I14). As one entrepreneur states "sustainability clashes with reality" (I1), as it would cost her too much time and money to operate her accommodation fully sustainable. Literature shows that time and money constraints are commonly named as barriers to more thorough implementation [37].

The CSR measures implemented in the accommodations typically are what is referred to as "low-hanging fruit"; measures that require little investment and effort, and easily earn themselves back moneywise [39]. One can think of utility usage reducing measures such as LED lights (I5, I6, I7, I8, I9, I10) and water reducing taps (I10). Other measures oft used are related to the prevention and reduction of waste, mostly food waste. For example, to have less food waste from breakfast, some ask the guest beforehand how much and what they wish for breakfast (I4 + I7). These are environmental measure found to be carried out often by hoteliers [36]. Others try to reduce the waste by eating leftovers themselves, or giving it as food for chickens (I1, I2). Arguably, the latter may be because of frugality, rather than sustainability, as it is "sinful to waste" (I2). This perfectly fits what Tzschenke et al. "inherited thriftiness" [36, p. 131]. Others stated that sometimes they had to waste "good food" due to hygiene regulations (I9, I15), or that they chose mono-packs as they believed that even though it used plastic, it prevented food waste (I1, I15). A more radical food waste prevention technique was used by one entrepreneur who refused to serve food in buffet form, as it led to too much food being thrown away (I17).

4.3 Values

Entrepreneurs were asked to select their top 5 values from a set of 17 value cards both for themselves personally, as for themselves as entrepreneur. From the 17 interviewed owner-managers nine persons created a value top 5 of most important values that was different for themselves personally and themselves as an entrepreneur. There are instances where respondents justify this difference by stating that as an entrepreneur, they have more obligations. They need to be more ambitious and focus on keeping the company financially healthy, as they "have bills to pay" (I7). Morrison and Teixeira [31] state that the economic survival and viability of the company is an important topic for these smaller (lifestyle) businesses. Eight persons indicated that they are the same person as an individual as an entrepreneur, and that their top 5 values is thus the same. As one entrepreneur states "what you see is what you get" (11).

From Tables 1 and 2 below, it can be seen that the personal values most often mentioned are pleasure, freedom and enjoying life. Pleasure and enjoying life are hedonic, whereas freedom falls under a gain goal frame. Looking at the personal values which score highest, it can be seen that freedom, pleasure and enjoying life are the most prominent. Even if one takes the subtotal of all biospheric items, it still doesn't become one of the most prominent self-reported values for individuals. Looking at the entrepreneur values, the items most often mentioned are freedom, pleasure and ambitious, with freedom scoring highest of the values, followed by enjoying life and pleasure. Enjoying life and pleasure are hedonic, and freedom is gain.

Table 1: Personal value top 5.

Personal value top 5		
Personal value	Score	Count
Freedom	54	14
Pleasure	45	15
Enjoying life	43	12
Ambitious	24	9
Helpfulness	20	7
Sub-total egoistic	173	54
Sub-total altruistic	47	17
Sub-total biospheric	36	14

Table 2: Entrepreneur value top 5.

Entrepreneur value top 5		
Entrepreneur value	Score	Count
Freedom	58	14
Enjoying life	36	10
Pleasure	33	13
Ambitious	31	11
Unity with nature	25	8
Sub-total egoistic	178	57
Sub-total altruistic	37	14
Sub-total biospheric	40	14

Comparing the two tables to each other, one can see that there is not a large difference between the ranking of the values according to personal or entrepreneur top five. Part of this may be understood as almost half of the sample saying they are the same person as an individual and as an entrepreneur. The biospheric value items score slightly higher with the entrepreneurial values. This might be explained because in tourism (under which hospitality and thus accommodation falls) nature is oft part of the product sold. What stands out from the explanation of the values by the respondent, be it for themselves personally, or for themselves as entrepreneur, is that most of them do not distinguish between the items that fall in the biospheric category. Even though the definitions of the terms were in the back, when picking one of the biospheric items the explanation of it overlaps with the meaning of the other items. In several interviews the respondent even said that to them these items were the same, or so closely related to each other that they found it hard to choose just one (I5, I14). If all the items in the biospheric category are taken together, they would move to position 4 in the personal values ranking, and even position 2 in the entrepreneur values ranking. The table further supports the claim that at least in this sample group, most owner-managers are not what is considered a sustainable entrepreneur, but rather entrepreneurs who undertake CSR measures in their accommodations to a varying degree. The entrepreneurs are typically motivated by self-enhancement values, though some self-transcendent values are of importance.

When combining motives, sustainability and values, profiles of the different entrepreneurs can be created. For space reasons, the author will present one entrepreneur who, compared to others in the sample, reported having implemented more CSR measures in his accommodation and one entrepreneur who reported fewer CSR measures implemented.

One of the interviewees reporting more, and more stringent CSR measures implemented in his accommodation is a 64-year old male hotel-restaurant owner-manager, falling in the passion for the industry category (I17). He is a 5th generation hotel owner, and it is a lifestyle he was used to from at home. He chose to become an owner-manager as it would give him more independence and freedom. He feels sustainability is a very broad concept, but to him it meant minimizing one's impact on the environment, and thinking about continuity. The measures undertaken by him are especially strong on the social dimension of sustainability; he offers his employees fixed contracts rather than temporary ones, he has used the same suppliers for his company since starting, and offers many vegetarian and vegan menu options. Furthermore, he is Green Key certified (highest level), avoids waste as much as possible by e.g. not offering buffets in the restaurant, uses local, sustainable, and/or fair-trade products as much as possible and renewable energy. His personal values are the same as those as an entrepreneur. In his top 5 he lists social justice (2nd place), equality (3rd place) and preventing pollution (5th place).

Not all accommodation entrepreneurs reported having implemented as many or high effort measures as the one described above. One entrepreneur with fewer CSR measures reported is presented here below.

This interviewee reporting fewer sustainability measures implemented than others in this research sample is a 43-year old female owner-manager of a hotel-restaurant who falls in the passion for the industry category (I9). She always wanted to own a hotel, and when moving within the country used that opportunity to realize her dream. She also appreciated the possibility of having a family with working from home. She defines sustainability as thinking about earth in the long term and not wasting things. Sustainability measures reported in her hotel are switching of lights when they are not needed, lowering the heating when possible, and installing LED lights. When part of her hotel was renovated, roof insulation was put in. She states herself that the sustainability measures she implements are out of cost saving, and that she evaluates possible measures using that as a criterion. She stated that the guest's expectations of luxury and the hotel building being very old are barriers to becoming more sustainable. She lists her personal values and entrepreneurial values to be different, both consist of self-enhancement values except for "helpfulness". This helpfulness value is stated by several interviewees to be part of the hospitality DNA, and needed to work in this industry.

Of course, there are many nuances in between the two ends of spectrum presented here. To exemplify; one 32-year old male entrepreneur with a passion for something else does focus on the wellbeing of his employees, and using local seasonal ingredients, but hasn't implemented many other more stringent CSR measures in his business.

5 CONCLUSIONS AND IMPLICATIONS

This paper reported on the motives and values of micro and small sized accommodation owner-managers to enter into (sustainable) entrepreneurship. Combining the motives for entering into the hospitality industry, their definition of sustainability and the measures implemented by them with their dominant individual and entrepreneurial values is a new avenue of research in this field. Though the sample size for this study limits the ability to make firm conclusions, it does give interesting insights deepening the knowledge in this field of research, and opens paths for further research.

The results of this research indicate that the reasons for entering into the accommodation industry is not just out of passion for it, or lifestyle reasons, but also by chance. This group of entrepreneurs was labelled as "accidental hoteliers". Notwithstanding, entrepreneurs in either category can fall on the higher or lower end of the spectrum when looking at number and type of CSR measures implemented in their accommodation.

When asking the interviewees to define sustainability several entrepreneurs sum up CSR measures that are or could be taken in their business, rather than giving an actual meaning of it. Furthermore, almost all respondents refer to the environmental side of sustainability, and do not recognize the social dimension as being part of it. Some of these entrepreneurs even undertook several measures which add value to the social side of sustainability, but it doesn't occur to them to mention these when either giving the definition, or explaining the measures undertaken by them.

When looking at values, the card game as used in this research is an innovative manner to understand the respondent's dominant values based on 16 values established to be significant for pro-environmental behaviour + 1 value from entrepreneurship theory, and used in questionnaire form. Overall, freedom, enjoying life and pleasure, all three self-enhancement values, rank highest in both the personal values, as well as the entrepreneurial values. When looking at the top 5 values listed of accommodation owner-managers with more and/or higher altruistic and biospheric values, we can recognize that these individuals also implement more and higher-effort sustainability measures in their business than others in this sample. Within the altruistic values, helpfulness and equality rank high. The helpfulness is seen as a trade needed when working in the hospitality industry, and equality was explained as having little power distance between themselves and their employees. Respondents found it hard to distinguish between the different biospheric values, and mostly explained them to belong to each other.

Though previous research has been carried out connecting reasons for entering the tourism industry to type and amount of sustainability measures implemented, this has not been connected to the persons' dominant values. This research implies that entrepreneurs listing more self-transcendent values in their top 5 personal and/or entrepreneurial values, also exhibited more and higher-effort sustainability measures implemented, and individuals with dominant self-enhancement values to have implemented fewer. Further (quantitative) research could prove the strength of this relation.

REFERENCES

[1] Walle, A.H., Business ethics and tourism: From micro to macro perspectives. *Tourism Management*, **16**(4), pp. 263–268, 1995.

[2] Steg, L., Values, norms, and intrinsic motivation to act proenvironmentally. *Annual Review of Environment and Resources*, **41**, pp. 277–292, 2016.

[3] Ateljevic, I. & Doorne, S., "Staying within the fence": Lifestyle entrepreneurship in tourism. *Journal of Sustainable Tourism*, **8**(5), pp. 378–392, 2000.

[4] Font, X., Garay, L. & Jones, S., Sustainability motivations and practices in small tourism enterprises in European protected areas. *Journal of Cleaner production*, **137**, pp. 1439–1448, 2016.

[5] Thomas, R., Shaw, G. & Page, S.J., Understanding small firms in tourism: A perspective on research trends and challenges. *Tourism Management*, **32**(5), pp. 963–976, 2011.

[6] UN, Goal 12: Ensure sustainable consumption and production patterns, n.d. www.un.org/sustainabledevelopment/sustainable-consumption-production/.

[7] Cohen, B. & Winn, M.I., Market imperfections, opportunity and sustainable entrepreneurship. *Journal of Business Venturing*, **22**(1), pp. 29–49, 2007.

[8] Schaltegger, S. & Wagner, M., Sustainable entrepreneurship and sustainability innovation: categories and interactions. *Business Strategy and the Environment*, **20**(4), pp. 222–237, 2011.

[9] World Travel & Tourism Council, Evaluation of job creation in G20 countries through maximising infrastructure capacity through technology that facilitates travel and integrates security, 2018.

[10] UNWTO, European Union tourism trends. UNWTO: Madrid, 2018.

[11] OECD, OECD Tourism trends and policies 2020. OECD, 2020.

[12] CBS Statline, Bedrijven; naar economische activiteit (SBI 2008, 2006-2010), 2012.

[13] UNWTO (United Nations World Tourism Organization), From Davos to Copenhagen and beyond: advancing tourism's response to climate change. UNWTO, 2009.

[14] MVO Nederland, Nut en noodzaak, 2015. http://mvonederland.nl/sector/toerisme. Accessed on: 11 Dec. 2015.

[15] Sloan, P., Legrand, W. & Chen, J.S., *Sustainability in the Hospitality Industry*, Butterworth-Heinemann: Boston, MA, 2009.

[16] Garay, L. & Font, X., Doing good to do well? Corporate social responsibility reasons, practices and impacts in small and medium accommodation enterprises. *International Journal of Hospitality Management*, **31**(2), pp. 329–337, 2012.

[17] Ahmad, S.Z., Entrepreneurship in the small and medium-sized hotel sector. *Current Issues in Tourism*, **18**(4), pp. 328–349, 2015.

[18] Lashley, C. & Rowson, B., Lifestyle businesses: Insights into Blackpool's hotel sector. *International Journal of Hospitality Management*, **29**(3), pp. 511–519, 2010.

[19] Fotiadis, A.K., Vassiliadis, C.A. & Rekleitis, P.D., Constraints and benefits of sustainable development: A case study based on the perceptions of small-hotel entrepreneurs in Greece. *Anatolia*, **24**(2), pp. 144–161, 2013.

[20] Drucker, P., *Innovation and Entrepreneurship: Principles and Practices*, HarperTrade: New York, 1985.

[21] Schwartz, S.H., Universals in the content and structure of values: Theoretical advances and empirical tests in 20 countries. *Advances in Experimental Social Psychology*, **25**(1), pp. 1–65, 1992.

[22] Schumpeter, J.A., The theory of economic development, translated by Redvers Opie. *Harvard: Economic Studies*, **46**, pp. 1600–0404, 1934.

[23] Skokic, V. & Morrison, A., Conceptions of tourism lifestyle entrepreneurship: Transition economy context. *Tourism Planning & Development*, **8**(2), pp. 157–169, 2011.

[24] Dewhurst, P. & Horobin, H., Small business owners. *The Management of Small Tourism and Hospitality Firms*, pp. 19–38, 1998.

[25] Tomassini, L., The co-creation of diverse values and paradigms in small values-based tourism firms. *Tourism Recreation Research*, **44**(3), pp. 359–369, 2019.

[26] Sarasvathy, S.D., Causation and effectuation: Toward a theoretical shift from economic inevitability to entrepreneurial contingency. *Academy of Management Review*, **26**(2), pp. 243–263, 2001.

[27] Steg, L., Bolderdijk, J.W., Keizer, K. & Perlaviciute, G., An integrated framework for encouraging pro-environmental behaviour: The role of values, situational factors and goals. *Journal of Environmental Psychology*, **38**, pp. 104–115, 2014.

[28] Teng, Y.M., Wu, K.S. & Liu, H.H., Integrating altruism and the theory of planned behavior to predict patronage intention of a green hotel. *Journal of Hospitality & Tourism Research*, **39**(3), pp. 299–315, 2015.

[29] DeCuir-Gunby, J.T., Marshall, P.L. & McCulloch, A.W., Developing and using a codebook for the analysis of interview data: An example from a professional development research project. *Field Methods*, **23**(2), pp. 136–155, 2011.

[30] Marcketti, S.B., Niehm, L.S. & Fuloria, R., An exploratory study of lifestyle entrepreneurship and its relationship to life quality. *Family and Consumer Sciences Research Journal*, **34**(3), pp. 241–259, 2006.

[31] Morrison, A. & Teixeira, R., Small business performance: a tourism sector focus. *Journal of Small Business and Enterprise Development*, 2004.

[32] Williams, C.C., The motives of off-the-books entrepreneurs: necessity-or opportunity-driven? *International Entrepreneurship and Management Journal*, **5**(2), p. 203, 2009.

[33] Melissen, F., Cavagnaro, E., Damen, M. & Düweke, A., Is the hotel industry prepared to face the challenge of sustainable development?. *Journal of Vacation Marketing*, **22**(3), pp. 227–238, 2016.

[34] Elkington, J. & Rowlands, I.H., Cannibals with forks: the triple bottom line of 21st century business. *Alternatives Journal*, **25**(4), p. 42, 1999.
[35] Alcott, B., Review: The conundrum: How scientific innovation, increased efficiency, and good intentions can make our energy and climate problems worse, David Owen, Riverhead Books, New York (2011), ISBN: 978-1-59448-561-9, 2013.
[36] Tzschentke, N.A., Kirk, D. & Lynch, P.A., Going green: Decisional factors in small hospitality operations. *International Journal of Hospitality Management*, **27**(1), pp. 126–133, 2008.
[37] Bohdanowicz, P., Environmental awareness and initiatives in the Swedish and Polish hotel industries—survey results. *International Journal of Hospitality Management*, **25**(4), pp. 662–682, 2006.
[38] Lepoutre, J. & Heene, A., Investigating the impact of firm size on small business social responsibility: A critical review. *Journal of Business Ethics*, **67**(3), pp. 257–273, 2006.
[39] Klewitz, J. & Hansen, E.G., Sustainability-oriented innovation of SMEs: a systematic review. *Journal of Cleaner Production*, **65**, pp. 57–75, 2014.

ILLUMINATING GREEN HOTELS: UNIVERSITY STUDENTS' REAL-WORLD EDUCATIONAL ACTIVITIES

PAULETTE HEBERT & GREGORY CLARE
Oklahoma State University, USA

ABSTRACT

Hotels can play an important role in tourists' experiences. Illumination is a critical component in hotel design. The objective of this study is to document appropriate educational activities which advocated the design of green (sustainable and healthy) hotels while specifically addressing relevant illumination issues, (i.e., reducing energy consumption through the use of Light Emitting Diodes (LEDs) and state-of-the-art lighting controls; and using "daylight harvesting") so as to prepare University students to understand and improve hotel designs of the future. This presentation will explore three very different, faculty-led, educational activities (Part 1, Part 2 and Part 3) related to green hotels, all conducted with the ultimate goal of understanding and improving tourist experiences: Part 1). Graduate student workshop at international conference: document hotel guest room temporary LED conversion and determine light spectra and light level measurement activity using spectrometers; Part 2). Undergraduate student in-class field study: document LED replacements in an on-campus hotel guest room's temporary conversion and light spectra and light measurement activity using spectrometers and light meters; Part 3). Six-week undergraduate student design studio class project to renovate lighting of an existing hotel such that it was sustainable. Use of energy-efficient light sources and lighting controls while minimizing watts/square feet were important considerations. Students utilized spectrometers and AutoCAD Enscape software to explore existing conditions or develop hotel renovation recommendations. N=61 students participated in the three activities. Students gained knowledge of "real-world", appropriate lighting design approaches for green hotels which could influence tourists' future experiences.

Keywords: lighting, hotel, energy efficient, LED, green, sustainable, tourism.

1 INTRODUCTION

A "green" hotel may be known as a(n) "sustainable," "eco-friendly" or "environmentally friendly" property [1]. The United States Environmental Protection Agency [2] defines sustainability in terms of "people, planet and prosperity". When considering sustainable tourism and green hotel illumination, the "people" reference may include assuring the health and safety of tourists at hotels (i.e., safety from slips and falls caused by inadequate light levels on stairs; freedom from eyestrain caused by inadequate light levels or glare; mitigation of circadian rhythm disruption associated with international travel/disruption of day/night cycle through the utilization of light level "doses" and light color spectra specification. For example, light spectra high in blue have been perceived as having a waking effect.) The "planet" reference could be addressed in hotels through the utilization of energy-efficient light sources; lighting controls which turn off or dim lighting when it is not needed. "Prosperity" could be found in the economy enjoyed by a hotel through reducing waste (i.e., turning off lights through daylight sensors and "harvesting daylight" when it is available to lower electric lights and resultant energy bills).

A green hotel incorporates environmental attributes such as conserving energy through its lighting system. A lighting system includes luminaires (light fixtures or fittings), light bulbs, wiring and lighting controls (dimmers, occupancy sensors etc.). Verma and Chandra [3] found that the guest rooms which utilized energy efficient light bulbs were perceived as most favorable by guests. The most energy efficient light bulb, whose use is trending and

WIT Transactions on Ecology and the Environment, Vol 248, © 2020 WIT Press
www.witpress.com, ISSN 1743-3541 (on-line)
doi:10.2495/ST200121

anticipated to become the hotel industry's standard, is the Light Emitting Diode (LED) [4]. LEDs consume less energy for light produced (lumens per wat) as compared to other light sources and need less replacement bulbs as they last 20,000 hours [5].

Some published hotel building standards intend to limit the amount of lighting energy consumed in different types of interior spaces to encourage sustainability. Large hotels incorporate guestrooms and function areas but may also include retail stores, museums and galleries, etc. The 2013 Building Energy Efficiency Standards [6] for Allowed Lighting Power Density Values prescribe the Watts/Square Foot allowed for various types of building spaces. These values range from 0.6 to over 2 watts per square foot for various space types within hotels. In some areas, these or similar standards may be adopted as law.

Also concerned with sustainability, the LEED Certification program, the USGBC (United States Green Building Council [7], a volunteer 3rd party organization, awards points to buildings which comply with its performance measures. Their hotel standard includes energy saving categories which appear to be applicable to lighting: "*Energy and Atmosphere*" (Prerequisite 2: "Minimum Energy Performance"; and "Credit 1: Optimize Energy Performance"); and "*Indoor Environmental Quality*" (Credit 6.1: Controllability of Systems – Lighting").

Instructor-guided, hotel lighting field studies have been shown to be effective in engaging students in learning "real world" lessons. Portable instruments, i.e. spectrometers have been found to be useful in capturing empirical data of existing conditions [8]. Nussbaumer has long advocated "evidence-based design" which consists of observation and documentation of existing conditions in order to inform future design projects [9]. Digital tools, such as AutoCAD, SketchUP and Enscape allow users to portray graphically complex images [10], and have proven useful for design students to convey their design solutions.

2 METHODS

A three-part methodology was utilized to provide learning experiences related to sustainable interior building illumination in hotels. The three activity sessions supported the overall goal of these educational activities, to improve tourists' experiences at hotels: Part 1). Graduate student workshop at an international lighting conference: light measurement activity documenting LED replacements in a conference hotel guest room's temporary conversion; Part 2). Undergraduate student in-class field study: light spectra and light level measurement activity documenting LED replacements in an on-campus hotel guest room's temporary conversion; Part 3). Six-week undergraduate student studio class project to design a lighting upgrade an existing hotel.

2.1 Part 1: Graduate student workshop

Graduate students attended a lecture in hotel meeting rooms about sustainable lighting design and circadian rhythm entrainment and then met in two designated hotel rooms for a hands-on data collection workshop. In hotel guestrooms A and B at an international conference, light color spectra and light level measurement activities were conducted in this graduate student field study. Light spectra measurement activities documented LED replacements in in a temporary conversion at a resort on the Western coast of Mexico. Room A utilized the existing hotel lighting. Room B was retrofitted with lightbulbs at the desk area in gooseneck lamps and the bedroom area in nightstand lamps. Prior to collecting lighting data, the students were trained in how to use and document readings a Lighting Passport ALP-01 spectrometer. Refer to Fig. 1 for image of students and instructor in retrofitted hotel room setup. Students worked in two-person teams and took turns using

Figure 1: View of retrofitted light bulb in nightstand.) (Left) and retrofitted light bulb at desk (Lower Right) in hotel at graduate student workshop with instructor and students.

Figure 2: Students-in-action at graduate student workshop. (Left and Middle) Students take spectral readings; (Right) Image of retrofitted desk lamp.

the instruments in Room A (existing) followed by Room B (retrofit) and documented readings between the two hotel rooms in their field notes. Refer to Fig. 2 for images of students-in-action in hotel room setup.

2.2 Part 2: Undergraduate student in-class field study

Light spectra and light level measurement activity documented LED replacements in an on-campus hotel guest room's temporary conversion. Undergraduate students met at a designated university hotel room during their regularly scheduled class period. Before conducting the field study, the students were provided with a tour of two adjoining hotel rooms. After the tour the students were trained in how to use and document readings from two instruments including an Extech HD450 light meter and a Lighting Passport ALP-01 spectrometer. Students worked in two-person teams and took turns using the instruments and documenting readings between the two hotel rooms. Room A used the existing LED hotel room lighting and Room B was retrofitted temporarily with various LED lightbulbs. The adjoining hotel rooms allowed the researchers and students to move freely between Room A

Figure 3: Images of existing lighting in Room A for undergraduate student lighting activity.

Figure 4: Images of retrofit lighting in Room B for undergraduate student lighting activity.

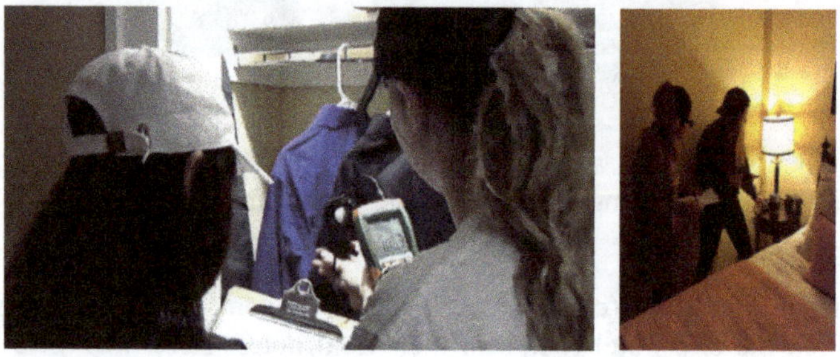

Figure 5: Students measure light levels in hotel room with spectrometer and light meter.

and B and discuss their readings as the student teams collected lighting data. Refer to Fig. 3 for images of existing Room A. Refer to Fig. 4 for images of retrofitted Room B. Refer to Fig. 5 for images of students-in-action during educational activity.

2.3 Part 3: Undergraduate student hotel design studio project

A six-week design studio class project was assigned which required lighting upgrades to an existing, historic, on-campus hotel. Students were given lectures and numerous reading assignments regarding green hotels and their appropriate illumination. The instructor issued a project statement outlining the course requirements for green lighting statements, plans, specifications and drawings for the hotel renovation. Shortly after the hotel project renovation was introduced, but after the students toured and documented existing conditions at the project hotel, the University transitioned to an "online only" classes format due to Covid 19 concerns. Thereafter, the instructor's guidance was delivered electronically via "announcements"; students' design progress was submitted electronically, bi-weekly to "discussion". Communication between instructor and students occurred on the Canvas online learning platform utilized by the University. Students posted written statements, questions and drawings to discussion board and received individual written feedback from the instructor.

Near the end of the six-week period, the students' design solutions were reviewed by the instructor and compared to the 2013 standards for Allowed Lighting Power Density Values (Watts/Square Foot). The following were determined to be the applicable spaces and allowed values for the current study:

Corridors, restrooms, stairs and support areas	0.6
Exhibits, Museums	2.0
Hotel function area	1.5
Retail merchandise sales	1.6
Other	0.6

3 RESULTS

3.1 Part 1: Graduate student workshop

Nine students were able to recognize spectral profile color variances when they measured the specific lightbulbs integrated into Rooms A and B. The retrofitted lightbulbs included two Lighting Science "Aware and Awake" (LS-75WE-BR40-9W-2349K-80 CRI bulbs which were retrofitted in the desk area in gooseneck lamps and two Lighting Science "Good Night" (FG-02263-A19-9W 2231K 70 CRI) lightbulbs placed in the nightstand table lamps. The light bulbs in the existing Room A and retrofitted Room B differed in characteristics. Refer to Table 1 for lighting retrofit bulb images and details regarding color spectra which were determined by students with the spectrometer.

Table 1: Graduate student workshop retrofitted lightbulbs description and spectral profile.

Description	Retrofit lightbulbs	
	Lighting science aware and awake	Lighting science good night
Lightbulb image		
Spectral profile	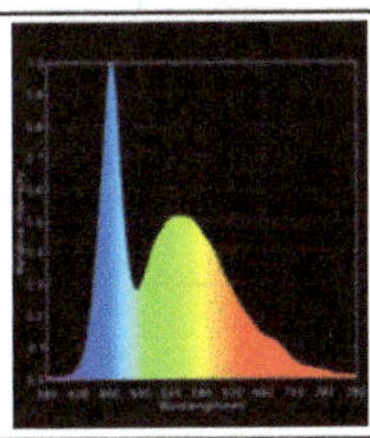	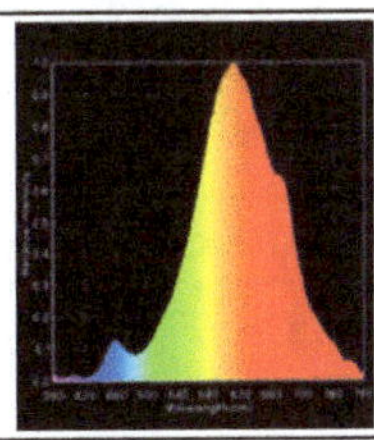

3.2 Undergraduate student in-class field study

A total of 36 students participated. They determined the varied spectral and illuminance differences between specific lightbulbs integrated in the retrofitted Room B. An example of differences noted included the Lighting Science "Aware and Awake" (LS-75WE-BR40, 9 Watt, 2349K, 80 CRI bulbs) which were retrofitted in the desk area in gooseneck lamps and Lighting Science "Good Night" (FG-02263-A19, 9 Watt, 2231K, 70 CRI) lightbulbs placed in the nightstand tables lamps. Refer to Table 2 for the color spectra for the retrofitted light bulbs as measured by students with spectrometer Refer to Fig. 5 for photographs of the hotel Room B lighting retrofit.

Table 2: Images and descriptions of retrofitted lightbulbs and their spectral profiles.

Description	Retrofit lightbulbs	
	Lighting science aware and awake	Lighting science good night
Lightbulb image		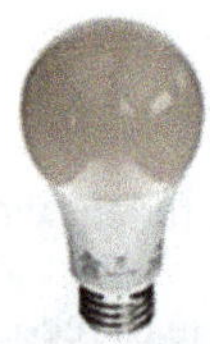
Spectral profile	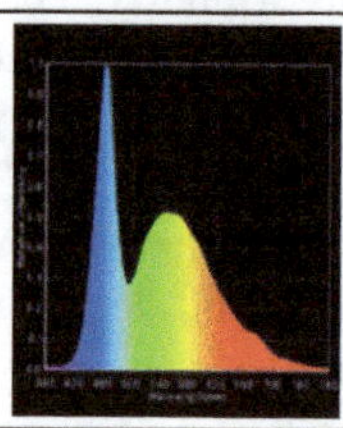	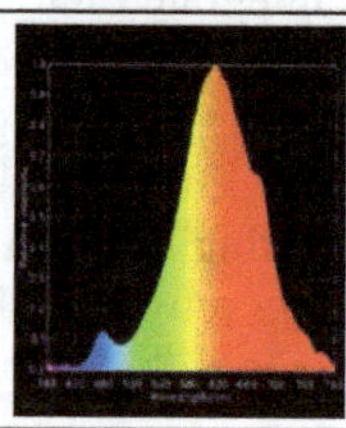

Refer to Tables 3 and 4 for Correlated Color Temperature (CCT), Color Rendering Index (CRI) and Illuminance (Lux and Footcandles) for Existing Room A and Retrofit Room B.

Table 3: Descriptive statistics of existing hotel room lightbulbs for the undergraduate student lighting workshop.

Room A (Existing)	Value			
	CCT	CRI(Ra)	Illuminance (Lux)	Illuminance (Footcandles)
Closet	2387 K	77	180 lux	16.7 fc
Desk	2553 K	79	188 lux	17.5 fc
Nightstand	2387 K	77	180 lux	16.7 fc
Bathroom	3127 K	83	182 lux	16.9 fc

Table 4: Descriptive statistics of retrofitted hotel room lightbulbs for the undergraduate student lighting workshop.

Room B (Retrofit)		Value			
	Light bulb	CCT	CRI(Ra)	Illuminance (Lux)	Illuminance (Footcandles)
Desk	Lighting Science Aware & Awake	2849 K	82	3261 lux	303.0 fc
Desk	General Electric Linksmart	2101 K	75	132 lux	12.2 fc
Nightstand	Lighting Science Good Night	7413 K	78	1854 lux	172.3 fc
Bathroom	General Electric Brightstick	2828 K	83	1443 lux	134.0 fc

3.3 Part 3: Undergraduate student hotel design studio project

Sixteen students participated in this educational activity and produced lighting statements outlining the sustainable lighting design solutions they employed; lighting plans with wiring and lighting control selections; watts per square foot calculations; lighting fixture schedules with light bulb indications; 3D renderings to depict the interior space solutions. AutoCAD, SketchUP and Enscape software were utilized. Refer to Figs 6, 7, 8, 9 and 10 for examples of students' deliverables for the hotel renovation project.

Figure 6: Example of student's lighting design solution for bar rendered in Enscape.

Figure 7: Example of student's lighting design solution for hallway rendered in Enscape.

Figure 8: Example of student's lighting design solution for guest room rendered in Enscape.

Hotel Lobby watts/sq.ft Recommendation: 1.1W/sq.ft.

Ketra D4R Retrofit Recessed Downlight
- 8 watts/fixture
- 9 fixtures total
- Total: 72 watts

Philips Hue
- 9.5 watts/fixture
- 10 bulbs total
- Total: 95 watts

X-Range LED Tape
- 1.0 watt/foot
- 36 feet total
- Total: 36 watts

Area of Lobby: 795 sq.ft.

Total Wattage
- 72 + 95 + 36 = 203 watts

Wattage/sq.ft.
- 203 watts/795 sq.ft. = 0.2553 watts/sq.ft.

Figure 9: Example of student's work showing hotel lobby power density recommendation for hotels compared to watts/square foot achieved in project solution.

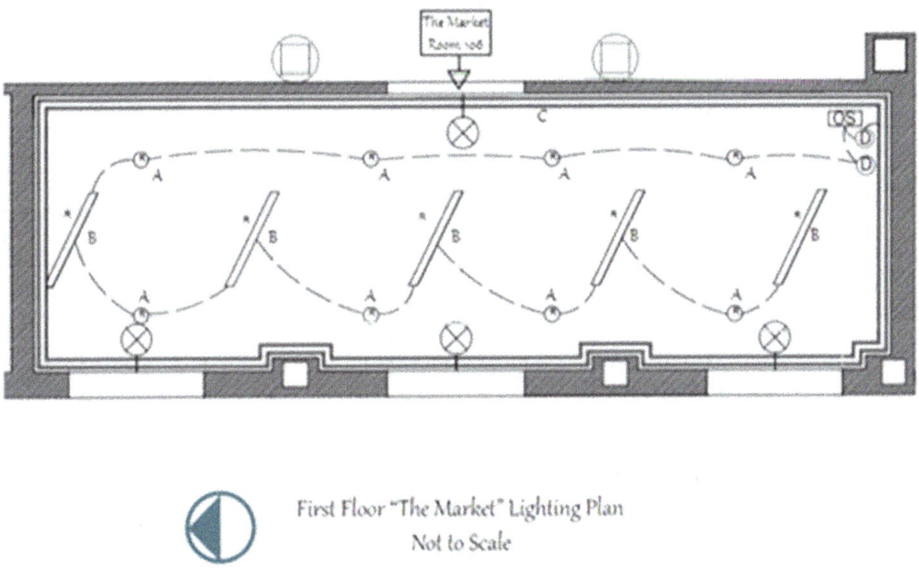

Figure 10: Example of a student's lighting plan for a retail store for the hotel design project with lighting controlled by an occupancy sensor (OS).

Upon review of the student projects, it was determined that the majority addressed the USGBC Credit Category, *Energy and Atmosphere*. Projects addressed the prerequisite, *Minimum Energy Performance* by providing evidence of compliance with the power density allowances for hotels. The student projects' reported power density values ranged from 0.16 to 6.7 watts/square foot. Refer to Table 5 for a breakdown of project compliance.

Table 5: Student project assessment of lighting USGBC credit category inclusions.

Student project assessment								
	Lighting USGBC credit category							
	Energy and Atmosphere				Indoor Environmental Quality			
	Prerequisite				Prerequisite			
	Minimum energy performance		Optimize energy performance		Controllability of systems – lighting			
Project #	W/sq ft complies w/ rec	Watts/ square feet	Light emitting diodes	Lumens/ watt	Occupancy sensor	Dimmer	Energy mgt. system	Daylight harvest/ sensor
1	Complies	0.31	x	NR	x	x	NR	NR
2	Complies	1.5	x	NR	x	x	NR	NR
3	Does not comply	2.55	x	NR	x	x	NR	NR
4	Complies	0.88	x	NR	x	x	NR	x
5	NR	NR	x	NR	x	x	NR	x
6	Complies	0.8	x	NR	x	NR	NR	NR
7	Does not comply	2.15–6.7	x	NR	x	x	NR	x
8	Does not comply	3.6	x	NR	x	x	NR	NR
9	Complies	1.2–1.4	x	NR	x	NR	NR	x
10	Complies	0.6	x	NR	x	x	NR	NR
11	Complies	0.55	x	NR	x	x	x	x
12	Complies	1.5	x	NR	x	x	NR	x
13	Complies	0.25	x	NR	x	x	x	x
14	Complies	1.5	x	NR	x	x	NR	NR
15	Partial compliance	1.04–1.53	x	NR	x	x	NR	NR
16	Complies	0.16	x	NR	x	x	x	x
Total								

NR =No report.

A majority of the students' design solutions (n=11, 68.75%) fully complied with the energy efficiency standards. Upon review of the students' projects it was determined that only half of the projects me the prerequisite *Optimize Energy Performance*. Refer to Table 5 for a breakdown of project compliance. While 100% of students' projects utilized energy efficient Light Emitting Diodes (LEDs) (n=16), none of the student projects specifically reported the associated lumens per watt (0, 0%) for LEDs.

Regarding the USGBC Credit Category, *Indoor Environmental Quality*, the majority of student projects were found to meet the applicable prerequisite, *Controllability of Systems – Lighting.* The majority of projects used multiple types of devices. All of the projects utilized occupancy sensors (100%, n=16). The vast majority of projects (n=14, 87.5%) utilized dimmer switches. Only three of the projects used energy management systems (18.75%). One half of the projects (n=8, 50%) used daylight harvesting which incorporated daylight sensors to turn off lighting fixtures near windows.

4 CONCLUSIONS

A substantial part of a tourist's experience may happen within a hotel. The sustainability of a hotel is determined by how it addresses people, prosperity and the planet. An important and highly visible component of a hotel is its lighting system. Illumination can be analyzed with respect to various factors including light color, energy consumption, lighting controls, quantity of light and quality of light. LEDs are considered to be sustainable light sources due to the amount of light they produce as compared to energy consumed and are recommended for sustainable hotels. LEDs are also sustainable because of their long lifetimes which decrease the need for maintenance, replacements, and landfill waste. Automatic controls, such as occupancy sensors, also are considered sustainable as they don't rely on humans to turn the lights off.

There are many possible educational approaches which could successfully facilitate students' learning with regards to the illumination of hotels. This study highlights three distinct strategies implemented by one instructor. Two of the approaches involved documenting and analyzing existing hotels' temporarily retrofitted LED lighting and the third strategy required students to design a new LED lighting system for an existing hotel. Various tools, including a spectrometer, a light meter and LEED certification program sustainability criteria, were used to determine lighting sustainability. These strategies increased participating students' awareness of sustainable lighting for hotels. The tourism industry would become more sustainable if hands-on educational experiences were implemented which were targeted towards the design and implementation of sustainable hotel lighting.

ACKNOWLEDGEMENTS

The authors wish to acknowledge the undergraduate and graduate students who participated in the educational activities contained herein, including but not limited to: Yang Pei, Madison Bachlor, Bethany Bailey, Haley Blakemore, Ashlyn Bone, Laney Braack, Madison Brook, Allison Castleberry, Katie Cooke Jennifer Galvin Cortes, Chandler Craven, Rebecca Dietzel, Amanda Dillow, Taylor Douglas, Jessica Duncan, Emma Edmonds, Jill Finnegan, Makaela Graham, Hannah Haines, Mariam Hamouch, Areanna Handy, Lane Hart, Sydney Hayes, Alicia Hernandez, Kylee House, Rannie Howard, Cayliah Jackson, Neema Kant, Violet Knowles, Skylar Littlechild, Ashyn Locklear, Kylie Lynch, Paige Mays, Sydney McAleb, Kathryn McMahan, Emmy Miller, Bethany Montgomery, Catherine Moore, Lauren Norwood, Maria Novak, Citalli Olivares-Mendez, Ashley Pokorny, Kara Rainey, Calleigh Sexton, Sophie Shelton, Becca Smith, Jessica Smith, Kyndall Smith, Kaelan Travis, Grace Wall, Brittany White and Madeline White.

REFERENCES

[1] Pizam, A., Green Hotels: A fad, ploy or a fact of life? *International Journal of Hospitality Management*, **28**(1), p. 1, 2009.

[2] United States Environmental Protection Agency, People, Prosperity and the Planet (P3), Student Design Competition. www.epa.gov/P3. Accessed on: 1 Jun. 2020.

[3] Verma, K. & Chandra, B., Hotel guest's perception and choice dynamics for green hotel attribute: A Mix method approach. *Indian Journal of Science and Technology*, **9**(5), pp. 1–9, 2016.
[4] Humphries, J., Six benefits of LED lighting for hotels in 2018, 2 May 2018. https://lodgingmagazine.com/six-benefits-led-lighting-hotels-2018/. Accessed on: 1 Jun. 2020.
[5] Peek, G., Johnston, J. & Hebert, P.R., Safe and sustainable lighting. Accessed on: 1 Jun. 2020. https://extension.okstate.edu/fact-sheets/safe-and-sustainable-lighting.html.
[6] Reference Ace. https://energycodeace.com/site/custom/public/reference-ace-2013/index.html#!Documents/57prescriptivemethodsfordetermininglightingpowerallowances.htm. Accessed on: 29 May 2020.
[7] U.S. Green Building Council, Practical strategies in Green Building: Hotels. www.usgbc.org/resources/practical-strategies-green-building-hotels. Accessed on: 26 May 2020.
[8] Hebert, P.R., Petrova, A., Clare, G. & Pei, Y., Exploring new light and color in and color instruments for multidisciplinary students' use: Empirical properties of existing building illumination and the effects of light on color samples. *Journal of Building Energy & Environment*, **1**(5), pp. 1–16, 2018. https://bee.kku.ac.th/wp-content/uploads/2018/06/bee-full-1-2018-Exploring-New-Light.pdf. Accessed on: 6 Jun. 2020.
[9] Nussbaumer, L., *Evidence-Based Design for Interior Designers*, Fairchild Books: New York, 2009.
[10] Eltanal, R. & Avelino, R., Architectural design course: performance innovative task output preference of 4th year BS-architecture students of the University of Science and Technology of Southern Philippines, Cagayan de Oro. *Open Journal of Social Sciences* **8**(2) pp. 1–11, 2020. DOI:10.4236/jss.2020.82005. www.scirp.org/html/5-1763154_98379.htm. Accessed on: 1 Jun. 2020.

APPLYING CARBON FOOTPRINT METHOD POSSIBILITIES TO THE SUSTAINABLE DEVELOPMENT OF SPORTS TOURISM

HRVOJE GROFELNIK[1], MARKO PERIĆ[1] & NICHOLAS WISE[2]
[1]University of Rijeka, Croatia
[2]Liverpool John Moores University, UK

ABSTRACT

Tourism generates diverse and complex economic, socio-cultural and environmental impacts on space. Tourism organizers and service providers tend to emphasize the positive effects, and often underestimate or neglect negative environmental impacts. The environmental impact of tourism is clearly visible in the transportation of tourists from their homes to their destinations and back, thus generating significant carbon dioxide (CO_2) emissions that directly participate in global warming. In order to determine the intensity of the impact of transport (for tourism purposes) on the environment, or more specifically on CO_2 emissions, the Carbon Footprint method (CF) is increasingly applied in modern scientific research. The possibilities of CF research in the form of case studies can be monitored in an increasing number and in various fields of research, confirming the value of the method. This review paper analyses the development and application of the CF method in the sustainable development of tourism and specifically sports tourism. By reviewing and analysing the relevant CF literature represented in the WoSCC database, this paper presents the basics of the scientific methodology of CF and its development and application from the aspect of sustainable development, with particular emphasis on the application of CF in measuring the environmental impact of tourism. The research points out the potential contribution of future CF research and complement methods in the overall picture of environmental burden. This should give an agenda and impetus for future research of sustainable development within the topic of the environmental burden generated by tourism.
Keywords: sustainable development, ecological footprint, carbon footprint, tourism, sport tourism.

1 INTRODUCTION

Tourism generates diverse and complex economic, socio-cultural and environmental impacts. In the context of sustainable development, in general, and tourism in particular, it is possible to distinguish two social groups with opposite positions from which they observe the environment and tourism. The first group are tourism organizers and service providers, who usually emphasize the positive effects and often underestimate or neglect the negative environmental impacts. The second group are people with high environmental awareness, is gaining influence, especially among tourists who have higher education attainment and higher income levels [1], [2]. Reconciling the views of the two groups is a development challenge for tourism in the near future, and an argumentative social debate based on scientifically learned knowledge is probably the way to bring the two groups closer together to formulate development strategies and models for environmentally sustainable tourism development. In the economic context of sustainable development, tourism (with about 10% of world GDP) is a key contributor of the modern global economic system [3]. In recent decades, the scientific and professional public has been increasingly following a trend that can be described by the rise of environmental and climate change research, which is also evident in the increasing number of scientific papers addressing the environmental impact of tourism [4]. Tourism has a direct or indirect impact on the environment through various physical interventions in space (infrastructure), generation of solid and liquid waste and gas emissions. Transportation of tourists from their homes to their destinations and back

www.witpress.com, ISSN 1743-3541 (on-line)
doi:10.2495/ST200131

generates significant carbon dioxide (CO_2, carbon) emissions that directly participate in contemporary anthropogenic global warming. In order to determine the intensity of the impact of transport (for tourism purposes) on the environment, or more specifically on carbon emissions, the Carbon Footprint method (CF) is increasingly applied in modern scientific research.

The aim of this paper is to review and synthesise papers that cover the issue of CF within the sustainable development of the tourism industry, and in particular with regard to the development of sports tourism which is proved to be one the most prosperous types of tourism [5], [6]. To accomplish this, we define the following research questions:

- Research Question 1: What is the state of the art in published tourism CF papers?
- Research Question 2: What approaches and methods or frameworks have been used to determine/measure the environmental burden of carbon emissions?

An overview of the CF research in tourism in this paper will systematize this growing area of scientific interest and point out the possibilities of complementing the overall picture of environmental burden with topics that have not yet been addressed and serve as an impetus for future research and complementing the image of sustainable development within the theme of environmental burden generated by tourism.

2 CONCEPTS AND BASIC FACTS OF THE SCIENTIFIC FIELD

The CF method is part of a broader methodological approach to determining the Ecological Footprint (EF) as one of the methodologies that have been developed over the last thirty years to determine current global or local load on space. In one of the early papers [7], EF is defined as a measurement tool that allows for determining the degree of utilization of a resource base and the possibility of assimilation of waste to a particular population on a particular surface. The methodology was designed to calculate EF by surface for a given population encompasses a wide range of human activities and their environmental impacts. Therefore, the EF methodology can be used integrally for the whole spatial system and its activities or partially. Partial use can be applied to isolated physical spatial units, but also to certain functional parts of spatial systems, such as for example individual activities or aspects of activities that can be explored as separate units in the context of sustainable development. In WoSCC, the first papers to determine the bioabsorption basis of CO2 as the basis for developing the CF method were published by Cienciala et al. in 1998 [8] while the CF method has been using the WoSCC database under this name since 2007 [9]. The possibilities of CF research in the form of case studies that can be tracked in increasing numbers in various fields of research to date confirm the value of the method This review paper analyses the development and application of the CF method in the sustainable development of tourism, and in particular sports tourism. The scientific public's interest in the CF method is increasing as global awareness of the environmental burden caused primarily by anthropogenic carbon emissions from the use of fossil fuels with specific effects on contemporary climate change.

The connection between sports, tourism and CF is multifaceted. Sport is a performance-oriented activity and is an important content of stay in a destination. Many tourists, when in a destination, decide to actively (as competitors) or passively (as spectators) participate in various sports activities. Additionally, sport increasingly becomes a tourist attraction [10] and the main motive for travelling to a certain tourist destination. This laid the foundations for the development of sports tourism and many authors discussed this phenomenon as being a unique interaction of places, activities, time and people [5], [10]–[12]. Today, it is argued that between 25% and 30% of the world's tourism economy is related to sport, that sports tourism is one of the fastest growing sectors in travel industry and shows no sign of slowing down [13], [14].

While there are numerous types of sports tourism, sport events are the most popular manifestation of sports tourism [15], [16]. Sport events, as well as sport activities in general, attract tourists (as participants or spectators) and could serve as a catalyst for tourism development if leveraged properly. Besides the impact on tourism and economy in general, they create diverse and intense socio-cultural and environmental impacts on a host community [17], [18]–[20]. While positive impacts are accentuated, negative environmental impacts in particular are often underestimated or neglected [17], [21]. Negative environmental event impacts such as environmental damage, increased amount of litter, waste, air pollution (e.g. transport related carbon dioxide emissions) and noise levels all happen within a destination.

3 METHODOLOGY

The study of CF in tourism and sport tourism is a relatively new area of academic research involving multiple disciplines, methodological approaches and topics. Based on this fact as well as on this study's objectives, systematic quantitative literature review is applied [22]–[24]. A systematic quantitative literature review follows a structured method to address particular research questions and aims for methodological transparency and reliability as well as systematic and comprehensive search on a specified topic. It may also incorporate quasi-statistical approaches in order to categorise, quantify, and identify trends in extant research thus assisting in highlighting research gaps and possible streams of future research [23], [25]–[27].

In conducting the review, a multistep process was used. First, a comprehensive review was conducted in December 2019 and January 2020 through searching the Web of Science Core Collection (WoSCC) database for publications that contain the term *Tourism* and *Carbon Footprint* in the title, abstract or keywords from the early publishing dates to December 2019. WoSCC have international coverage, include top rated publications and are therefore largest databases of peer-reviewed literature. In view of the scope of this review paper, a selection of publications by the WoSCC database was started according to the Topic criterion with the key word *tourism* and 75,362 papers were found in database. After using the *and function* for subjoin *Carbon Footprint* in the WoSCC database a total of 179 publications were found. Initial cursory analysis of selected publications revealed that many of the selected papers were not research papers and/or that they deal with the CF concept in a trivial and marginal way. These papers were therefore excluded from further analysis.

However, when it comes to publications focused on sport tourism (searching for the terms *tourism* and *carbon footprint* and *sport* in the title, abstract or keywords) the sample was reduced to only seven papers. Two out of seven papers had not considered the CF of sport tourism like the object of the study. For this reason, using these particular papers' list of references as a guide, further search of the database is conducted, and additional 8 papers is found to be relevant (although they do not have the term *Tourism* and *Carbon Footprint* in the title, abstract or keywords). The final sample consisted of 13 publications closely related to the CF in sport tourism, all of which papers published in scientific journals.

Due to the relatively small number of articles included in the study, the authors stored, coded, and categorized papers manually. In order to classify papers into categories, direct extraction and/or inductive interpretation methods were implemented (see Thomson et al. [26]). Frequency of appearance of particular categories was searched within the selected papers. Inductive method in interpretations allowed new findings to emerge through the data.

The categorisation process was iterative to ensure consistency of the categorisation of articles in the sample.

4 RESULTS AND DISCUSSION

4.1 Review of papers with topic on carbon footprint in tourism

Analysing the dynamics of CF research papers in tourism at WoSCC database it was found that the first paper using words CF was published by in 2007 [28], and after sporadic work in the initial phase of developing the CF method after 2010 a positive upward trend on an annual basis is visible (Fig. 1). Because CF method derive from the EF concept in the initial phase of developing CF concept research, various authors used name of EF.

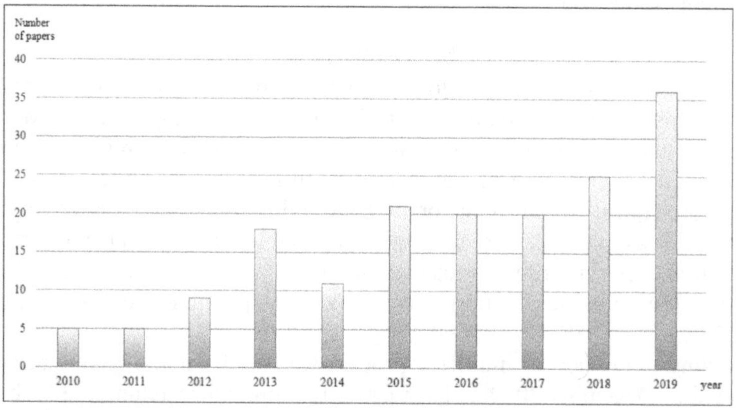

Figure 1: Dynamics of tourism CF research papers referred in WoSCC 2010–2019.

As mentioned above, the search of the WoSCC database resulted in 179 publications, most of which (139) are articles published in scientific journals. In addition, there are 17 papers published in conference proceedings, six reviews, five book chapters, two books and six other scientific publications.

Looking at the spatial interest of the researcher (Table 1), it is evident that there is no homogeneity at the world level. Several large and developed countries such as Australia, China, the UK and Sweden are leading the research field of CF in tourism. Notably, there is a lack of research from large and developed countries such as USA and Japan. This suggests that Australia and the UK have a broader base of the researchers, while in the rest of the countries/universities have individual researches on tourism CF.

By analysing the WoSCC database for CF method in tourism over two decades, only four scientific journals stands out by a number of papers published. These four journals have published more than ten papers each and they account for 38% of the total number of papers covered by this analysis. Journal with the most published papers on CF of tourism is *Journal of Sustainable Tourism* (22 articles, Q1), then *Journal of Cleaner Production* (18 articles, Q1), then *Sustainability* (16 articles, Q2) and *Tourism Management* (12 articles, Q1).

A content analysis of the titles of the identified papers reveal the following terms that are repeated more than five times: *Tourism* with 98 citations, *Carbon* with 91, *Carbon footprint*

Table 1: Spatial distribution of papers with the most active authors in the field.

State	University with more than three published papers	Author with more than three published papers
Australia	University of Queensland (9 papers)	Dolnicar, S. (5 papers)
	Griffith University (7 papers)	Becken, S. (4 papers)
	James Cook University (4 papers)	-
China	Chinese Academy of Sciences (8 papers)	-
	Institute of Geographic Sciences Natural Resources Research CAS (4 papers)	-
	Sichuan University (4 papers)	-
UK	Bournemouth University (15 papers)	Filimonau, V. (13 papers), Robbins, D. (5 papers), Dickinson, J. (4 papers)
Sweden	Linnaeus University (7 papers)	Hall, C.M. (4 papers)
	Lund University (5 papers)	Gossling, S. (9 papers)
Hong Kong	Hong Kong Polytechnic University (6 papers)	-
Taiwan	National Cheng Kung University (6 papers)	Sun, Y.Y. (7 papers)
Canada	University of Waterloo (4 papers)	Scott, D. (4 papers)
Netherlands	Breda University of Applied Sciences (4 papers)	-

with 42, *Carbon emission/s* 14, and *Transportation* with 6. From the above it is evident that Carbon footprint in tourism is linked with transport as the largest emitter of CO_2 related to today's global tourism.

Further, a focus on methodology for CF calculation was considered. In-depth analysis of papers revealed that many of the selected papers were not research papers. More precisely, 37 papers are purely theoretical/conceptual studies (that study the basics of CF in tourism), 43 papers are empirical using different data sources, and 99 papers are dominantly theoretical with marginal empirical part. For this reason, only 43 empirical papers are taken for further analysis and others were therefore excluded.

Among the analysed papers 43 empirical papers were singled out of which 25 used top-down approach to collect the data with secondary sources of data in which the most used and important are methods: (1) Tourism Satellite Account (TSA), e.g. [29]; (2) Input-output model (IO), e.g. [30]; (3) Life cycle assessment (LCA), e.g. [31]. In the context of tourism CF methods data collection the main challenge is their capability to keep track of indirect carbon emissions associated with the products and services hidden in the supply chain. To resolve this problem many researchers used input–output analysis (IO). IO, however, suffers from a limited scope of application – primarily at macro scales, such as national industrial sectors, national economies, national tourism industries and national tourist markets [32]. To reduce the problem of indirect carbon emissions, over time a method of LCA has evolved in which both the direct and the indirect carbon emissions for energy and material inputs of a goods and services over its life cycle are included.

Additionally, 18 papers used bottom-up approaches, using primary sources of data obtained from surveys of tourists or economic operators [33], institutions managing an area

[34] or conducted in direct contact with transport companies [35]. The bottom-up approach used primary data sources to provide accurate and more secure outcomes. However, the weakness of such an approach is in the organisation and duration of data collection, spatial and partial system coverage of such research. It would take many case studies to get the overall picture, which would require a lot of researchers and time for data collection and processing.

4.2 Review of papers with topic on carbon footprint of sports tourism

Our search resulted in 13 scientific journals directly related to the CF in sport tourism (Table 2). What is evident from the Table 2 is the variety of examined sports and events as well as approaches to measure CF. While the type of sporting events range from local community based sporting events (e.g. ice hockey [36] or university athletics [37]) to global mega events (e.g. FIFA World Cup or Olympic Games [38], [39]). It was determined that most studies focused on internationally popular sports like soccer, American football, running, cycling and athletics.

Considering the authors and their affiliations, it is not surprise that these events were held in large and developed countries like the UK, USA or Canada. Despite being a large and competitive country, and the fact that many authors who studied CF were from China, when it comes to sporting events, there were not any studies from China. Additionally, in most of the previous studies, data were collected from event attendees, both competitors (athletes or teams) and spectators. While in first studies, authors mostly applied IO analysis, recent studies favoured LCA and other innovative methods.

5 CONCLUSION

The development of tourism in modern times is increasingly diverse, its environmental impact is growing, and the developmental dynamics of environmental tourism research can be clearly observed in the number of papers referred in the WoSCC database. In the last three decades, the development of the EF and specifically CF method has followed the contemporary development of the awareness of the scientific and general social public about the need for environmental protection and the need for sustainable development. In this context, in order to answer the research questions raised at the outset, this paper reviewed the development and state of the art of the CF method in tourism and sport tourism in particular.

Providing a systematization of this growing area of scientific interest, this paper presents the basics of the CF scientific methodology. CF method initially derives from the EF concept, and in the initial phase of CF research various authors used name of EF. Expanding the search on other databases, conducting more detailed content analysis of other publications in the database, and inclusion of additional terms such as *Ecological Footprint* or *Carbon Emissions* in our search would certainly result in other relevant papers. This can guide future studies in this fast-growing area of research.

The spatial scope of the analysed papers was different (from local to global level) as well as applied approaches and methodologies. Two basic frameworks for calculating CF in tourism have been identified: a top-down analysis dominantly using secondary data sources and a bottom-up analysis using primary data sources. Despite the advance in the development of CF determination methods, the problem of indirect CF in general and indirect CF of tourism is still not fully resolved. The dominance of theoretical/conceptual scientific papers among the published papers has been observed, and in order to further development of the CF field, it is necessary to make steps towards studying more cases at the local and regional

Table 2: Relevant CF sport tourism papers referred in WoSCC database.

Author	Sport/Event	Country	Method/s
Collins et al. [40]	Football (soccer), FA Cup Final	UK	- Input-output (IO) - Ecological footprint (EF)
Jones [41]	Rally, World Rally Championship	UK	- Input-output (IO) - Tourism satellite accounting - Interviews
Collins et al. [38]	Multiple sports, Summer Olympic Games and FIFA World Cup	UK	- Input-output (IO) - Ecological footprint (EF) - Survey method - Information from the event sponsors
Chard and Mallen [36]	Ice hockey, local matches	Canada	- Interviews - Carbon footprint calculators (CarbonZero and PlanetAir)
Collins et al. [42]	Cycling, UK Stages of Tour de France	UK	- Input-output (IO) - Ecological footprint (EF) - Survey method - Carbon footprint (CF)
Krugell and Saayman [43]	Marathon, world race	South Africa	- Survey method - Regression analysis
Dolf and Teehan [44]	Multiple sports, University sports events	Canada	- Life Cycle Assessment (LCA) method - Survey method
Edwards et al. [37]	Athletics, University sports events	USA	- Survey method - Student observers - Information from event organizers - Life cycle assessments (LCA + SimaPro LCA software)
Scrucca et al. [45]	Orienteering, World Championship	Italy	- Carbon footprint (CF) analysis - Information from organizers, athletes and accommodation - Authors proposes a method of summary indicator
Pereira et al. [39]	Football (soccer), FIFA World Cups	Multiple states	- Facility Location Problem (FLP) - mathematical modelling method for determining the best location with lowest transportation CF
Triantafyllidis et al. [46]	American football, Collegiate sport events	USA	- Survey method - Independent t-Test - ANOVA analysis
Wicker [6]	Multiple sports	Germany	- Survey method - Regression analysis
Pereira et al. [47]	Football (soccer), English Premier League matches	UK	- Carbon footprint (CF) assessment method developed by the UK's DEFRA - Questionnaire

levels. This is especially important if policies are going to be realised locally so that development strategies at the local and regional levels can embrace CF from the bottom-up. In the analysis, we noticed the need to improve empirical CF researches with field case studies that would upgrade the overall picture of measuring the environmental load. Furthermore, the overall picture of the pressures put on the environment will be much upgraded by conducting a greater number of empirical studies of CF tourism and sports tourism. Based on greater number of works based on primary source data in future it is possible to create a wider base that will help in creating sustainable development policies on a scientific basis.

Regarding the limitations of this study, we searched only WoSCC database. It is also possible that there were papers in the WoSCC database that deal with CF in tourism and/or sport tourism but they did not use the terms *tourism*, *carbon footprint* and *sport* in the title, abstract and keywords. But this selection of papers indexed in WoSCC are well-regarded for peer-review rigour and assessing these sources guarantees that the authors of this study are assessing papers that are reaching the widest possible audience.

ACKNOWLEDGEMENT

This paper was funded under the project line ZIP UNIRI of the University of Rijeka, for the project ZIP-UNIRI-116-2-19.

REFERENCES

[1] Liu, K.S., Hsueh, S.L. & Chen, H.Y., Relationship between environmental education, environmental attitudes, and behavioral intentions toward ecolodging. *Open House International*, **43**(2), pp. 5–12, 2018.

[2] Dodds, R., Graci, S.R. & Holmes, M., Does the tourist care? A comparison of tourists in Koh Phi Phi, Thailand and Gili Trawangan, Indonesia. *Journal of Sustainable Tourism*, **18**(2), pp. 207–222, 2010. https://doi.org/10.1080/09669580903215162.

[3] Petcu, M.A. & Cismasu, I.D., Tourism: A contribution to the sustainable economic growth. *Quality-Access to Success*, **20**, pp. 460–466, 2019.

[4] Velasco-Muñoz, J.F., Aznar-Sánchez, J.A., Belmonte-Ureña, L.J. & López-Serrano, M.J., Advances in water use efficiency in agriculture: A bibliometric analysis. *Water*, **10**, pp. 377, 2018. https://doi.org/10.3390/w10040377.

[5] Weed, M. & Bull, C., *Sports Tourism: Participants, Policy and Providers*, 2nd ed., Elsevier: Oxford, 2009. https://doi.org/10.4324/9780080942117.

[6] Wicker, P., The carbon footprint of active sport participants. *Sport Management Review*, **22**(4), pp. 513–526, 2019.

[7] Wackernagel, M. & Rees, W.E., *Our Ecological Footprint: Reducing Human Impact on the Earth*, New Society Publishers: Philadelphia, 160 pp., 1996.

[8] Cienciala, E., Running, S.W., Lindroth, A., Grelle, A. & Ryan, M.G., Analysis of carbon and water fluxes from the NOPEX boreal forest: Comparison of measurements with FOREST-BGC simulations. *Journal of Hydrology*, **212**(1–4), pp. 62–78, 1998. https://doi.org/10.1016/S0022-1694(98)00202-9.

[9] Christen, K., The carbon footprint of transportation fuels. *Environmental Science and Technology*, **41**(19), pp. 6636–6636, 2007.

[10] Hinch, T.D. & Higham, J., Sport tourism: A framework for research. *International Journal of Tourism Research*, **3**, pp. 45–58, 2001. https://doi.org/10.1002/1522-1970(200101/02)3:1<45::AID-JTR243>3.0.CO;2-A.

[11] Gammon, S. & Robinson, T., Sport and tourism: a conceptual framework. *Journal of Sport Tourism*, **4**(3), pp. 11–18, 1997. https://doi.org/10.1080/14775080306236.

[12] Sobry, C., Liu, X. & Li, J., Sport tourism: Contribution to a definition and categorization. *Acta Turistica*, **28**(1), pp. 7–26, 2016.

[13] Hoye, R. & Parent, M.M., *The SAGE Handbook of Sport Management*. SAGE: Thousand Oaks, 2016, ISBN 978-1-4739-5923-1.

[14] Parker, R., The remarkable growth of sport tourism. Human Kinetics, 2019. https://humankinetics.me/2019/04/04/the-remarkable-growth-of-sport-tourism/. Accessed on: 11 Feb. 2020.

[15] Getz, D. & Page, S.J., Progress and prospects for event tourism research. *Tourism Management*, **52**, pp. 593–631, 2016. https://doi.org/10.1016/j.tourman.2015.03.007.

[16] Roche, S., Spake, D.F. & Joseph, M., A model of sporting event tourism as economic development. *Sport, Business and Management: An International Journal*, **3**(2), pp. 147–157, 2013. https://doi.org/10.1108/20426781311325078.

[17] Ahmed, T.S.A.A., A triple bottom line analysis of the impacts of the Hail International Rally in Saudi Arabia. *Managing Sport and Leisure*, **22**(4), pp. 276–309, 2017. https://doi.org/10.1080/23750472.2018.1465841.

[18] Perić, M., Đurkin, J. & Wise, N., Leveraging small-scale sport events: Challenges of organising, delivering and managing sustainable outcomes in rural communities, the case of Gorski kotar, Croatia. *Sustainability*, **8**(12), pp. 1337, 2016. https://doi.org/10.3390/su8121337.

[19] Schulenkorf, N. & Edwards, D., Maximizing positive social impacts: Strategies for sustaining and leveraging the benefits of intercommunity sport events in divided societies. *Journal of Sport Management*, **26**(5), pp. 379–390, 2012. https://doi.org/10.1123/jsm.26.5.379.

[20] Wise, N., Outlining triple bottom line contexts in urban tourism regeneration. *Cities*, 53, pp. 30–34, 2016. https://doi.org/10.1016/j.cities.2016.01.003.

[21] Perić, M., Estimating the perceived socio-economic impacts of hosting large-scale sport tourism events. *Social Sciences*, 7(10), pp. 176, 2018. https://doi.org/10.3390/socsci7100176.

[22] Cooper, H., Hedges, L. & Valentine, J., *The Handbook of Research Synthesis and Meta-Analysis*, 2nd ed., Russell Sage Foundation: New York, 2009.

[23] Pickering, C. & Byrne, J., The benefits of publishing systematic quantitative literature reviews for PhD candidates and other early career researchers. *Higher Education Research and Development*, **33**(3), pp. 534–548, 2014, https://doi.org/10.1080/07294360.2013.841651.

[24] Wakefield, A., Synthesising the literature as part of a literature review. *Nursing Standard*, **29**(29), pp. 44–51, 2015. https://doi.org/10.7748/ns.29.29.44.e8957.

[25] Onwuegbuzie, A. & Frels, R., *Seven Steps to a Comprehensive Literature Review: A Multimodal and Cultural Approach*, SAGE, 2016.

[26] Thomson, A., Cuskelly, G., Toohey, K., Kennelly, M., Burton, P. & Fredline, L., Sport event legacy: A systematic quantitative review of literature. *Sport Management Review*, **22**(3), pp. 295–321, 2019. https://doi.org/10.1016/j.smr.2018.06.011.

[27] Whittemore, R. & Knafl, K., The integrative review: Updated methodology. *Methodological Issues in Nursing Research*, **52**(5), pp. 546–553, 2005. https://doi.org/10.1111/j.1365-2648.2005.03621.x.

[28] Fawkes, S., Carbon dioxide emissions resulting from space tourism. *Journal of the British Interplanetary Society*, **60**, pp. 409–413, 2007.

[29] Wu, D.C.G., Liu, J.Y., Song, H.Y., Liu, A.Y. & Fu, H., Developing a web-based regional tourism satellite account (TSA) information system. *Tourism Economics*, **25**(1), pp. 67–84, 2019. https://doi.org/10.1177/1354816618792446.

[30] Sun, Y.-Y., Decomposition of tourism greenhouse gas emissions: Revealing the dynamics between tourism economic growth, technological efficiency, and carbon emissions. *Tourism Management*, **55**, pp. 326–336, 2016. https://doi.org/10.1016/j.tourman.2016.02.014.

[31] Sharp, H., Grundius, J. & Heinonen, J., Carbon footprint of inbound tourism to Iceland: A Consumption-based life-cycle assessment including direct and indirect emissions. *Sustainability*, **8**(11), pp. 1147, 2016. https://doi.org/10.3390/su8111147.

[32] Filimonau, V., Dickinson, J.E., Robbins, D. & Reddy, M.V., A critical review of methods for tourism climate change appraisal: Life cycle assessment as a new approach. *Journal of Sustainable Tourism*, **19**(3), pp. 301–324, 2011. https://doi.org/10.1080/09669582.2010.527345.

[33] Michailidou, A.V., Vlachokostas, C. & Moussiopoulos, N., A methodology to assess the overall environmental pressure attributed to tourism areas: A combined approach for typical all-sized hotels in Chalkidiki, Greece. *Ecological Indicators*, **50**, pp. 108–119, 2015. https://doi.org/10.1016/j.ecolind.2014.11.006.

[34] Monz, C., D'Antonio, A., Lawson, S., Barber, J. & Newman, P., The ecological implications of visitor transportation in parks and protected areas: Examples from research in US National Parks. *Journal of Transport Geography*, **51**, pp. 27–35, 2016. https://doi.org/10.1016/j.jtrangeo.2015.11.003.

[35] Grofelnik, H., A regular annual sea transport carbon footprint for the islands of Cres and Lošinj. *Croatian Geographical Bulletin*, **77**(2), pp. 73–83, 2015. https://doi.org/10.21861/HGG.2015.77.02.04.

[36] Chard, C. & Mallen, C., Examining the linkages between automobile use and carbon impacts of community-based ice hockey. *Sport Management Review*, **15**(4), pp. 476–484, 2012. https://doi.org/10.1016/j.smr.2012.02.002.

[37] Edwards, L., Knight, J., Handler, R., Abraham, J. & Blowers, P., The methodology and results of using life cycle assessment to measure and reduce the greenhouse gas emissions footprint of "Major Events" at the University of Arizona. *The International Journal of Life Cycle Assessment*, **21**(4), pp. 536–554, 2016.

[38] Collins, A., Jones, C. & Munday, M., Assessing the environmental impacts of mega sporting events: Two options? *Tourism Management*, **30**(6), pp. 828–837, 2009. https://doi.org/10.1016/j.tourman.2008.12.006.

[39] Pereira, R.P.T., Camara, M.V.O., Ribeiro, G.M., & Filimonau, V., Applying the facility location problem model for selection of more climate benign mega sporting event hosts: A case of the FIFA World Cups. *Journal of Cleaner Production*, **159**, pp. 147–157, 2017.

[40] Collins, A., Flynn, A., Munday, M. & Roberts, A., Assessing the environmental consequences of major sporting events: The 2003/04 FA Cup Final. *Urban Studies*, **44**, pp. 457–476, 2007. https://doi.org/10.1080/00420980601131878.

[41] Jones, C., Assessing the impact of a major sporting event: The role of environmental accounting. *Tourism Economics*, **14**(2), pp. 343–360, 2008. https://doi.org/10.5367/000000008784460382.

[42] Collins, A., Munday, M. & Roberts, A., Environmental consequences of tourism consumption at major events: an analysis of the UK stages of the 2007 Tour de France. *Journal of Travel Research*, **51**, pp. 577–590, 2012. https://doi.org/10.1177/0047287511434113.

[43] Krugell, W. & Saayman, M., Running a greener race: Willingness-to-pay evidence from the Old Mutual Two Oceans Marathon in South Africa. *South African Journal for Research in Sport, Physical Education and Recreation*, **35**(1), pp. 53–70, 2013.

[44] Dolf, M. & Teehan, P., Reducing the carbon footprint of spectator and team travel at the University of British Columbia's varsity sports events. *Sport Management Review*, **18**(2), pp. 244–255, 2015. https://doi.org/10.1016/j.smr.2014.06.003.

[45] Scrucca, F., Severi, C., Galvan, N. & Brunori, A., A new method to assess the sustainability performance of events: Application to the 2014 World Orienteering Championship. *Environmental Impact Assessment Review*, **56**, pp. 1–11, 2016. https://doi.org/10.1016/j.eiar.2015.08.002.

[46] Triantafyllidis, S., Ries, R.J. & Kaplanidou, K., Carbon dioxide emissions of spectators' transportation in collegiate sporting events: comparing on-campus and off-campus stadium locations. *Sustainability*, **10**, 241, 2018. https://doi.org/10.3390/su10010241.
[47] Pereira, R.P.T., Filimonau, V. & Ribeiro, G.M., Score a goal for climate: Assessing the carbon footprint of travel patterns of the English Premier League clubs. *Journal of Cleaner Production*, **227**, 167–177, 2019.

REMOTE ISLAND TOURISM: A CASE STUDY IN FIJI

HIROSHI KATO
Centre for Advanced Tourism Studies, Hokkaido University, Japan

ABSTRACT

The purpose of this paper is to make clear that "Remote island tourism" is one of the frontiers for South Pacific nations. Fiji plays a vital role in this paper as the main case study for pushing forward the idea. In order to confirm the existing situation and to evaluate the potential of remote island tourism, the author conducted field surveys to remote islands, including the Lau archipelago, Lomaiviti archipelago, and Kadavu island in Fiji. They sit on the eastern side or southern side of the Fiji main island, Viti Levu. It was confirmed that signs of long periods of cultural interchanges between different races clearly remain on remote islands in forms such as architectural styles. Based on the observations obtained from the field surveys, this study concludes that "Remote island tourism" is one of the frontiers for South Pacific nations. This study identified that the lifestyle of the remote islanders can be one of the tourist resources for people who live in urbanised areas. Regarding ways to deal with associated challenges, this paper suggests that "Remote island tourism" should be pursued in step with developments of social basic needs including transportation in the remote islands.

Keywords: remote, Viti Levu., Lau, Kadavu, Bau, Levuka, Lakeba, roof, resource, jetty.

1 INTRODUCTION

Fiji has been reputed to be a leading resort destination in the South Pacific. Foreign inbound tourists numbered 755,000 in 2016 [1]. "Blue sea and white beach" have long been placed at the centre of the destination images of Fiji, and tourism development has been promoted on a full scale since around end of the 20th century. Neighbouring nations, however, have been pursuing similar destination images. In addition, Fiji's tourist destinations are unevenly distributed on the western coast of Viti Levu. Therefore, it is reasonable for Fiji to make efforts to establish a stratified destination image and to expand tourist destinations nationwide. Little specific information, however, has been gained on what kinds of local resources exist and how they can be used for tourism promotion beside the western coast of Viti Levu.

It is generally believed that several thousand years ago Papuans firstly migrated to Fiji and that they were followed by Melanesians and then Polynesians [2]. Times advanced to the modern and in the early 1800s, sandalwood trade brought Europeans to Levuka because this fragrant wood was discovered north of Ovalau Island [3]. With this kind of historical background, it is easy to understand that Fijian culture has been influenced by those of several races and folks.

The 320 islands which comprise Fiji are spread over an area of 709,700 sq. km [4]. The total population of Fiji is estimated at about 850,000, and national capital Suva lies on Viti Levu, the largest island in the nation. Together with the second largest island called Vanua Levu, more than 90% of the national total population live on the two islands.

The purpose of this paper is to make clear that "Remote island tourism" is one of the frontiers for South Pacific nations. "Remote islands" in this paper exclude largest-class islands or socioeconomically most advanced island in a nation such as the capital island. In the case of Fiji, Viti Levu and Vanua Levu are not object islands of this paper.

This study concludes that "Remote island tourism" is one of the frontiers for South Pacific nations, it also identified that lifestyle of the remote islanders can be one of the tourist resources for people who live in urbanised areas. Regarding ways to deal with associated

www.witpress.com, ISSN 1743-3541 (on-line)
doi:10.2495/ST200141

challenges, this paper suggests that "Remote island tourism" should be pursued in step with developments of social basic needs including transportation in the remote islands.

2 FIELD VISITS FOR REMOTE ISLAND TOURISM

2.1 General

In this paper, Case Study Method is adopted taking into consideration of both research accuracy and efficiency. As discussed in the previous section, there are more than 300 islands in Fiji, which are spread over about 700,000 sq. km (Fig. 1). Distances between ports and islands are shown in Table 1.

Figure 1: Fiji Islands. *(Source: Kato [5].)*

Table 1: Distances between ports. *(Source: Hydrographic office, Fiji Navy [6].)*

PORT	SUVA	TOTOYA	MATUKU	MOALA	LAKEBA	CICIA	V.BALAV	LEVUKA	VUNISEA
SUVA		113	99	95	167	138	165	52	59
TOTOYA			27	29	76	83	116	106	118
MATUKU				46	103	104	141	102	97
MOALA					74	64	101	87	115
LAKEBA						45	55	143	187
CICIA							39	103	165
V.BALAVU								125	205
LEVUKA									94

Unit: nautical miles.

Object islands for the case study are carefully evaluated and selected. Because existing tourist destinations are unevenly distributed on the western coast of Viti Levu. Therefore, it is appropriate to focus on rather eastward or southward from the viewpoint of information gathering. Object islands for the case study are selected taking into consideration of sizes (land area and population) and geographical distribution: Lau archipelago, Lomaiviti archipelago, and Kadavu Island.

2.2 Lau archipelago

Lau archipelago lies between Melanesian Fiji and Polynesian Tonga, and is composed of about sixty islands and islets, about thirty are inhabited. This archipelago has a population of about 10,000 in total [7]. In July 2016 Arthur visited six islands among the Lau archipelago aboard MV Iloilovatu, which Fiji Government owned and operated for disaster relief or transportation of the government supplies for island people (560 GRT, 56 m in length, 17 crew members). Itinerary of field visits by the MV Iloilovatu including land area and population of the visited island is shown on Table 2. Arthur spent all nights aboard the ship during this field visits. MV Iloilovatu had been used as a training vessel of a Japanese high school for fishery education.

Table 2: Itinerary of field visits by Iloilovatu.

July 2016		ACTIVITIES
10	S	Departure from Suva Port
11	M	Moala Is. (65 sq.km and 1,384 population)
12	T	Totoya Is. (28 sq.km and 538 population)
13	W	Totoya Is.
14	R	Matuku Is. (57 sq.km and 672 population)
15	F	Matuku Is.
16	Sa	Lakeba Is. (59.5 sq.km and 1,751 population)
17	S	Lakeba Is.
18	M	Lakeba Is.
19	T	Vanua Balavu Is. (57 sq.km and 1,799 population)
20	W	Vanua Balavu Is.
21	R	Arrive at Cicia Is. (34 sq.km and 1,024 population)
22	F	Arrive at Cicia Is.
23	Sa	Arrival at Suva Port

2.2.1 Lakeba island

Lakeba island was an important meeting place between Fijians and Tongans throughout their early history. Tonga is geographically nearer than Viti Levu from the Southern Lau islands. In addition, the trade wind blows from the east to the west. Tongan influence has left its remarkable marks on the Lau soil [4].

Not only religious people but also warriors came to the Lakeba island together with Tongan culture. The first Christian missionaries to Fiji landed at Lakeba in 1830 and Ma'af, Tongan chief established himself at Lomaloma, Northern Lau island [4]. Tongan influence can still be found in various fields including architectural style. In Fiji architectural styles of most houses are characterised in square-shaped ends. Architectural styles of Lauan houses are, however, tend to be rounded, following the Tongan practice (Fig. 2). This Tongan architectural style with a rounded roof can be seen not only for public buildings (Fig. 3) but also for private houses.

Figure 2: General view of Tubou village, Lakeba island.

Figure 3: Roof structure of a church, Tubou village.

With an area of 54 sq. km and a population (1976) of 2067 in eight villages, Lakeba is the largest of the Lau islands. Tubou is the traditional chiefly village and is the seat of the traditional paramount chief, Tui Nayau [4].

Modern accommodations are not available in Lakeba island. However, there are two homestays in this island. Arthur was invited to one of the homestays in Tubou village to look at the interior (Fig. 4(a)). Structural frameworks of the rounded roof are amazing when seen from inside. Beams and pillars under the roof are decorated with white shells. Houses are standing around Village Green (open space) where is the place being used for socialisation by villagers. Villagc Grccn is well maintained.

Cargo–passenger boats irregularly call the island once or twice a month and supply daily necessities for island life. The boats do not have fixed shipping schedule because they cannot make profits by shipping operation. Therefore, service ship and port of call are decided monthly by a bid, and the government fills a deficiency,

There is an airfield in Lakeba island. A small plane with less than 10 seats provides services between the island and Nausori airport near the capital Suva. However, air service is only once a week. If you fly to the island by a plane, you must stay the village at least seven nights. In return, you can enjoy the village life, and understand how friendly dogs are (Fig. 4(b)).

(a) (b)

Figure 4: Tubou village, Lakeba island. (a) Inside of homestay, Lakeba island; and (b) Dogs siesta, Lakeba island.

2.2.2 Other islands

Matuku Is has a land area of about 28.5 sq. km. It has seven villages and a population in 1976 of 1044 [4]. The jetty is set at geographical centre of the island [8]. A young villager said to Arthur, "A road was constructed at the back yard of the primary school in around 2000. This road was supposed to connect the center of the village with the Jetty. The road has been unused because there were no cars in the village and gradually disappeared, becoming bush, and ultimately abandoned. This island is fully protected by coral reef, and lagoon is shallow. Arthur made a sail round the Matuku Island by a small fibre boat. Water inside the lagoon was extremely crystal. The seabed covered with coral was visible even aboard the boat. Villagers in Matuku Island are living with close relationship with animals. Pigs are left free to walk around any places in the village even at school campus. They enjoy strolling with a family and are getting fat as large as a calf" (Fig. 5(a)).

(a) (b)

Figure 5: Island life. (a) Pig family, Matuku Island; and (b) Inside of community centre, Cicia Island.

(Vanua Balavu Is.) is the third largest island in the Lau archipelago. In 1855, Ma'af acquired sovereignty over northern Lau and established Lomaloma as the capital of the Lau Confederacy [9]. Tongan cultural influences still can be seen in the form of the architectural style in Vanua Balavu island; many houses and public buildings with rounded roof stand in Lomaloma village. Lomaloma Jetty is serviced fortnightly under the Government Shipping Franchise Scheme. Vanuabalavu Airport has a 762 m runway, and services with regular flights to/from Nausori near the capital Suva. A couple of sailing boats mooring near the mouth of Lomaloma Bay were observed from the deck of the MV Iloilovatu when she departed the Lomaloma Jetty.

(CICIA Is.) has an area of 34 sq. km with five villages. It has a population of about 1,000. There is an airfield with a runway of 762 m. The runway is not paved but a grass field. Once a week (on Tuesday) a small aircraft connects the island with the Nausori Airport. Coastal road with gravel surface runs encircling the island. This island is serviced fortnightly under the Government Shipping Franchise Scheme. The Tongan cultural influences are still marked in the island life such as in the form of the architectural style; many houses or public buildings in the village have round shaped roof (Fig. 5(b)). Other island lifestyles may be seen during walking tour around the island. In Arthur's case an old villager was making copra fibre in the part of his garden.

2.3 Lomaiviti archipelago

Lomaiviti archipelago is a group of islands between Viti Levu and Lau archipelago. Ovalau, 13 km long and 11 km wide, is the principal island of this islands group. Census population of the Ovalau island is 8,360 in 2007.

2.3.1 Ovalau island

Levuka is the socioeconomic centre of this island (Fig. 6(b)). Moreover, Levuka was the first national capital of FIJI, and was firstly inscribed as a UNESCO world heritage site from the South Pacific islands. Unlike Lauan villages Levuka is basically European atmosphere town.

Levuka started to gain prominence in the early 1800s when European traders paid attention to sandalwood or beche-de-mer. In addition, a small whaling settlement had been built at this town by 1830. Both Levuka's geographical location in the centre of Fiji islands

(a) (b)

Figure 6: Levuka, Ovalau Island. (a) Colonial style architecture, Beach Street; and (b) Levuka town overlook.

and existence of sheltered harbour with two entrances/channels have contributed to establish Levuka as Fiji's commercial centre.

In 1871 Levuka became the capital of Kingdom of Fiji. After 1874 when Deed of Cession was signed, Levuka served as the Fiji's capital until 1882. Levuka has been a busy port town even after the title being the national capital was removed. Cotton trade has been booming at Levuka throughout in the 1870s, and copra entered a trading market as a new commodity.

In June 2013 Levuka Historical Port Town was listed as UNESCO World Heritage Site (cultural) because "It is a rare example of a late colonial port town, which illustrates the cultural hybridity of non-settler communities in the Pacific, with an urban plan that merges local settlement traditions with colonial standards" [10].

Near the harbour, small scale buildings with colonial style are lined along the beach (Fig. 6(a)). Every Saturday morning people gather at an open-air market held in front of the lined shops. Lovoni people also come down with agricultural products to sell from the mountain and buy daily necessities on their return home. Within a walking distance from the port entrance, Levuka village with traditional settings lies facing a creek. Levuka is no more busy port because of limited hinterland and lack of special products except canned fish. Sometimes, however, a few sailing boats mooring at the calm basin can be seen.

2.4 Viti Levu's adjacent islands

2.4.1 Bau Island

Bau island lies about 10 km north of Nausori airport and is separated from the main island Viti Levu by 300 m long shallow waters. The highest point of the island is 15 m above sea-level and its land area is only eight hectares.

Despite its modest size, Bau rose to prominence in the mid-1800s and became Fiji's dominant power. Bau is the chiefly home of Bauan people, and home and stronghold of the great King Cakobau, who made his people leaders of Fiji. To come from the Bau is a status symbol for Fijian. In its heyday, the island boasted about three thousand inhabitants and it has maintained its influence in politics through to modern Fiji.

In 1852 Cakobau became Chief of Bau and Kingdom of Fiji was established by Cakobau in 1871. Deed of Cession to Great Britain was signed at Levuka in 10 October 1874. Cakobau took leadership of this deal.

It should be noted that Ma'af, a chief and warrior from Tonga was one of the signers of this important political document. However, foreign influence and cultural exchange with them are obvious in landscape of this island.

At the field visit to Bau Island in 2017, Author was invited to the chief's residence to look inside. Usually Fijian architecture is composed of strait lines, and roofs extend upward. However, roofs of the chief's resident is circular at the ends. The chief's wife introduced herself saying, "I am related with Tonga" (Fig. 7(a)).

Village Green (open space) is situated at the central place of the island and well maintained as is the case in the Lakeba Island. The chiefly mausoleum stood on the knoll and with typical Fijian architectural style (Fig. 7(b)). Both two buildings of different architectural style stood closely with each other.

(a)

(b)

Figure 7: Various architectural styles in Bau Island. (a) Interior of the Chief's house; and (b) Fiji's traditional style architecture.

Currently the Bau Island is not completely open to tourists or visitors. Invitation is required to tourists or visitors who want to land the Bau Island because of limited land space and limited usable resources. However, tourists or visitors can enjoy village landscape and villagers lifestyle freely, encircling the island aboard fibre boats.

2.5 Kadavu Island

In Kadavu Island there are two jetties, Kavala Jetty and Vunisea Jetty. Through site visits it was found that they are new type of jetties from viewpoint of tourism. At 11 pm of 31 May 2017, Lomaiviti Princess (5,864 GRT, Ro-Ro/Passenger ship) departed from Muwaiwalu Wharf, Suva where the Author embarked. The next morning, she approached the Kavala Jetty, first port of call of in the Kadavu Island. When Arthur turned back on the deck, he recognised many fibre boats coming somewhere or other toward the Ro-Ro ship. Around 9 am of 1 June 2017, she arrived at the Kavala Jetty (Fig. 8(a)).

Even before some passengers and cars have completed disembarkation there, small fibre boats came one after another to the Jetty. The boats were filled with passengers and baggage. On the backup area, tents were expanded by villagers for selling meals, oranges, talo and so on. The backup area has changed from vacant grass field to a festival marketplace.

Traditionary, jetties in Fiji do not have a backup area, but the Kavala Jetty has. Ratu Sir Kamisese Mara states as follows:

"All candidates were invited to set out their platforms in the Fiji Times, and mine was quite simple – communications in Lau and Lomaiviti. I would work for the widening and deepening of entry passages through the reefs, small jetties with storage sheds and rudimentary feeder roads, more radio and telephone stations, a freight subsidy, and a school for boat building" [11].

At the Kavala Jetty, a spatial reclamation area has been prepared at the back yard for parking of vehicles and for an open market (Fig. 8(b)). Passengers both aboard and incoming/outgoing can take a break there and enjoy talking each other for a while. Most of the island jetties have been built solely as a ship docking point. No other functions have been taken into consideration. Progressive jetties should be planned as waterfront space, not a point, which accommodates a living function as well as a logistic function.

(a) (b)

Figure 8: Island style, Kadavu Island. (a) Just after the ferry has arrived; and (b) Jetty is filled with festival atmosphere.

Kadavu is about 48 km in length and varies in breadth from about 365 m to 13 km. Its area is 411 sq. km and an island population is about 10,000. Land transport is very limited in the Kadavu Island, and Kadavu airfield is not accessible from Kavala Village by land. Vunisea is the central place of Kadavu Island. Copra, fruits, and good timber abounds on the south and south-east slopes [4].

2.6 Summary of field visits

Through the field visits it was revealed that existing service level of the transportation system in Fiji remote islands was very poor and constituted one of the most serious hindrance to the development of remote island tourism. Improvement of transportation system is a must for tourism development in remote islands.

Regarding the tourist resources, each village has its own unique history and heritage. In general sense, traces of cultural influences and exchanges among different folks/races can be a potential tourist resource in remote islands in Fiji. Fig. 9 shows selected impressive scenes for the author on several remote islands.

Figure 9: Selected scenes on the remote islands. (a) Disappeared tracks, Matuku Island; (b) Grass runway, Cicia airfield; (c) Copra fibre, Cicia Island; and (d) Kavala Jetty, Kadavu Island.

3 STRATEGIES FOR INCREASING TOURISM

Improvement of transportation infrastructure should be given the highest priority in order to materialize the remote islands tourism in Fiji. It is not an easy task to improve transportation infrastructure in a short period of time since there are about one hundred inhabited islands in Fiji's. In addition, the government budget is limited, too. Realistic ways must be adopted to materialize the goal.

"Staged development strategy" can be recommended in transportation development works to support the realization of the remote islands tourism in Fiji. It would be too optimistic to consider that upgrading the transportation infrastructure can be realised in all remote islands and number of incoming tourists would be increased simultaneously among all remote islands. Therefore, priority shall be given to a specific island group. If efficiency and effectiveness of the investment are taken into consideration, it would be reasonable to consider that the priority goes to the island group which is nearest to the national marketplace, Suva city.

The first stage of the works is to upgrade service level of the mother port in the main island. Muwaiwalu Wharf at Suva Port, which is separated from the main wharf and about 1 km away from the city centre, has been playing as a mother port function. However, the Muwaiwalu wharf has only 122.6 m in length and 48.8 m wide, hence the terminal is too small to accommodate all passengers, cargos, and vehicles of interisland shipping. In addition, it is old and deteriorated. Therefore, a new interisland terminal should be built somewhere, perhaps on the eastern coast of Viti Leve to serve for the Lau archipelago and Lomai Viti archipelago. However, the mother port function for Kadavu Island should remain at Muwaiwalu wharf in Suva Port, As Suva Port is faced to the South, most of the interisland ships must take a circuitous route when they destine to the Eastern islands. This is another reason why an additional interisland terminal should be built at a new site.

The second stage is to introduce high speed passenger boats between the new terminal and Lomai Viti archipelago to formulate one day trip sphere from Suva. Residents of Levuka say as follows.

> "The introduction of fast, comfortable boats that cruise at 14 knots and accommodate 50 passengers would also boost the economy. Direct runs from Suva to Levuka on Mondays, Wednesdays, and Fridays with runs in the other direction on Tuesdays, Thursdays, and Saturdays would assist inns and restaurants. Travellers might want to visit other nearby islands, such as Cagilai, Leluvia, Lost Island, and Naigani, which would be easy with small boats" [3].

Not only island residents, but also cruise passengers want high speed passenger boats to be introduced. Cruise ships usually arrive at ports early in the morning and depart a harbour late afternoon or in the evening. Thus, cruise passengers have 7–8 hours only to make an optional tour during a stay at Port.

The third stage is to connect Lau archipelago with Lomai Viti archipelago or other islands including Taveuni island and neighbouring nations. At this stage planners are advised to make a close watch at the socioeconomic moves and changes.

4 CONCLUSIONS

"Blue sea and white beach" have long been placed at the centre of the destination images of Fiji. Through the field visits to Lau archipelago, Lomai Viti archipelago, and Kadavu island, it was found that "culture" is one of the important local resources of remote islands in Fiji. One of the important elements of the tourism is "a displacement outside the usual environment". "Remote island" just meets with this criterion because in modern society people tend to live in urbanised areas. Time passes slower in remote islands than in urbanised areas. Therefore, past events (history) are felt clearer for islanders than city dwellers. Unique lifestyle still exists in remote islands in Fiji. The unusual environment is felt fresh for city dwellers. Lifestyle of islanders can be one of the tourist resources for people who live in urbanised areas. it is concluded that this research made clear that "Remote island tourism" is one of the new frontiers for Fiji. Improvement of transportation infrastructure should be given the highest priority in order to materialize the remote islands tourism in Fiji. "Staged development strategy" can be recommended in transportation development works to support the realization of the remote islands tourism in Fiji.

The first stage of the works is to upgrade service level of the mother port in the main island. A new interisland terminal should be developed on the eastern coast of Viti Leve to serve for the Lau archipelago and Lomai Viti archipelago. The second stage is to introduce high speed passenger boats between the new terminal and Lomai Viti archipelago to formulate one day trip sphere from Suva. The third stage is to connect Lau archipelago with Lomai Viti archipelago or other islands including Taveuni island and neighbouring nations. At this stage planners are advised to make a close watch at the socioeconomic moves and changes. It is suggested that "Remote island tourism" should be pursued in step with developments of social basic needs.

From discussion and consideration above, it is concluded that this research made clear that "Remote island tourism" is one of the new frontiers for South Pacific island nations.

ACKNOWLEDGEMENTS

This research work has been supported in part by Ministry of Infrastructure and Transport FIJI and Japan International Cooperation Agency (JICA). The contents of this paper are the sole responsibility of the author.

REFERENCES

[1] Kato, H., *Kotu infra kara kannko wo kanngaeru-Kannko sozogaku heno Challenge*, Center for Advanced Tourism Studies (CATS), Hokkaido University: Sapporo, pp. 319–325, 2017. (In Japanese.)

[2] Carter, J., *Fiji Handbook: Business and Travel Guide*, Pacific Publications: Sydney and New York, pp. 7–22, 1980.

[3] People of Levuka, Levuka: Living Heritage, Institute of Pacific Studies, University of the South Pacific and Levuka Historical and Cultural Society, pp. 96–97, pp. 136–137, 2001.

[4] Douglas, N. & Douglas, N., *Fiji Handbook: Business and Travel Guide*, Pacific Publications: Sydney and New York, pp. 9–17, pp. 210–213, 1987.

[5] Kato, H. (ed.), *Jetties and Landings in Fiji for Public Transport*, Ministry of Infrastructure and Transport Fiji, Japan International Cooperation Agency (JICA): Suva, 2017.

[6] Hydrographic Office, Fiji Navy, Fiji nautical almanac 2015, Hydrographic office, Fiji Navy: Suva, Fiji, 2014.

[7] Lau Islands, Wikipedia. https://en.wikipedia.org/wiki/Lau_Islands. Accessed on: 3 Mar. 2020.

[8] Kato, H., *Completion Report of H. KATO's Trip to Lau Group Island*, Ministry of Infrastructure and Transport Fiji and Japan International Cooperation Agency: Suva, 2017.

[9] Lomaloma, Wikipedia. https://en.wikipedia.org/wiki/Lomaloma. Accessed on: 3 Mar. 2020.

[10] Levuka Historical Port Town, UNESCO World Heritage Centre. http://whc.unesco.org/en/list/1399/. Accessed on: 10 May 2014.

[11] Mara, K., *The Pacific Way: A Memoir*, University of Hawai'i Press: Honolulu, 1997.

REWILDING URBAN AREAS TO MAINTAIN ECOSYSTEM SERVICES AND ECOLOGICAL TOURISM: THE CASE OF SALER'S DEVESA, VALÈNCIA, SPAIN

JOSÉ LUIS MIRALLES I GARCIA & FELIPE MARTÍNEZ LLORENS
Department of Urbanisme, Universitat Politècnica de València, Spain

ABSTRACT

The Saler's Devesa is a case of a failed tourist project to urbanize a historical protected forest inside the metropolitan area of València in Spain. In the 1970s, the government started actions to urbanize this area. In fact, all the public works and some buildings were built. However, in the 1980s the new democratic government decided to rewild the area and "deconstruct" the urban development works that had already been built. The case allows knowing the process of rewilding and the difficulties to do it. On the other hand, currently in developed countries, metropolitan areas simultaneously present a stabilized population and a great surface extension. In this situation, the sustainability of these urban areas has focused on managing the city already consolidated by renovation or regeneration processes and rewilding open spaces, developing green infrastructures. Because of this, we focused our attention on researching one of the few cases that exists on rewilding urban areas. The green infrastructure concept has been used in the case of La Devesa to maintain and increase ecosystem services. In addition, these actions allow promotion of ecological tourism inside the area. The research question is "What should be the idea of society's progress when touristic urban development growth is no longer possible or desirable?" The article analyses the historical process of rewilding and rethinking the idea of progress in the current situation.

Keywords: rewilding cities, green infrastructures, ecological tourism, rewilding cost, metropolitan areas.

1 INTRODUCTION

The case of "La Devesa de l'Albufera" or Saler's Devesa is an exceptional case of urban area rewilding. "La Devesa" is a sand strip between the Albufera Lake (fresh water) and the Mediterranean Sea (salt water) close to Valencia City (Spain). This territory was occupied by a dense Mediterranean forest and, since medieval age, was public property, for a long time such as royal lands and after such as part of natural heritage of Valencia City Council. Even this land was historically protected because of Valencia City Council, in the 19th century and about the first half of the 20th century, couldn't use it for other uses different of natural forest.

In the 1960s, as part of public policy of Spanish state for touristic development, it was elaborated an urbanistic plan to build a new tourist town inside the forest after selling public property. Of course, this urbanization was going to destroy the forest and its ecosystem. Valencia City Council paid the cost of public works to urbanize the area and sold the plots to build. In fact the public works to urbanization was completely executed and also a part of buildings.

In the 1970s, social values changed along with the political system that changed from Franco's dictatorship to democracy in 1978. The first democratic government stopped the urbanization process and started the process to rewilding the area in accordance with their political programme.

This case has been analysed by Martínez Llorens in his PhD research [1] directed by Dr Josep Luís Miralles. This article shows a part of this research and use that to review or rethink the concept of progress in currently situation of metropolitan areas.

WIT Transactions on Ecology and the Environment, Vol 248, © 2020 WIT Press
www.witpress.com, ISSN 1743-3541 (on-line)
doi:10.2495/ST200151

Therefore this article is structured in four sections plus this introduction. Section 2 shows the general trends about metropolitan areas and the introduction of green infrastructure concept. In addition, the main characteristics of the case of "La Devesa" such as case of urban touristic development. Section 3 shows the methodology and data, specifically about the process of rewilding and it economic cost. Section 4 includes the results and it discussion. And finally, Section 5 shows the conclusions.

2 FRAMEWORK AND LITERATURE REVIEW

2.1 Previous considerations

The case of "La Devesa" is extraordinary because is a case of de-urbanization of an already urbanized land. In fact, it is a case of forest-ecosystem rewilding with a hard destruction by a building process for a touristic development. The objective now is to identify how our society can make the most of this experience to solve current problems or achieve new goals, particularly the sustainability of urban touristic development or urban development in general.

To achieve this goal we will focus our attention on three aspects:

- General evolution of metropolitan areas that concentrate the main social and economic activities especially as response of economic system to financial crisis of 2008.
- The use of green infrastructure as way to regenerate urban and metropolitan areas on sustainability conditions.
- The case of "La Devesa" as an example of the rewilding process allowing knowledge of the economic costs.

2.2 Metropolitan trends

According to different studies and various authors such as Vheeler [2], Aguilera et al. [3], Feria Toribio and Martínez Bernabeu [4] or Miralles i Garcia [5], etc., metropolitan areas in developed countries show a trend, especially since the 2008 economic crisis, with two main characteristics:

- Stabilised population or low increase;
- Very significant increase of extension or surface. That is, increase of towns next to central areas with commuter travels and consequently increase of surface extension.

This situation is sign that, social and economic activities have been increased more in metropolitan areas than other territorial areas such as rural areas or small–medium cities, of course, before the coronavirus crisis.

In addition, this trend implies the need of focus the attention on urban renovation as a system to improve or change urban uses in order to satisfy new urban demands. Of course, this thinking is coherent with sustainability urban goals against the historical trend of transformation rural landscapes into urban areas.

Fig. 1 shows the current extension of the metropolitan area of València in Spain. In it you can see the Albufera Lake and the "La Devesa del Saler" that is the land band between the lake and the Mediterranean Sea. You can see that these spaces, protected today, are inside the metropolitan area of València. Today these spaces are part of protected areas system because of different causes (natural ecosystems, agricultural areas, natural or historical heritages, etc.). Albufera Lake is a special case of wetland transformation to a semi-artificial

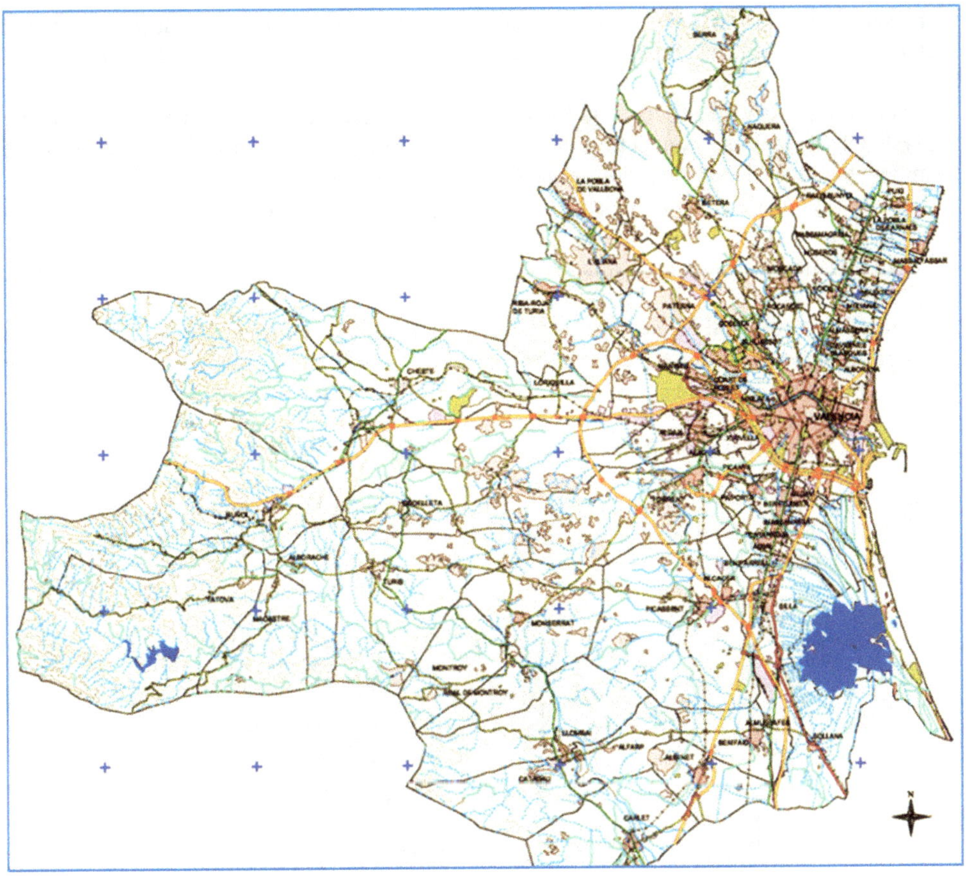

Figure 1: Metropolitan area of València. *(Source: UDR F.Eiximenis.)*

ecosystem [6]. In addition, there are a cities system and a large list of infrastructures, especially transports infrastructures, mostly designed from the only point of view of it function to serve cities but not to respect, maintain or promote the ecosystem services of land's sectors crossed by them.

As a consequence, within the framework of general trends in metropolitan areas, stabilized/low growth population, the sustainable urban growth to satisfy new demands of urban uses only is possible by urban renovation processes. If these new demands or urban uses are satisfied by building new rural areas instead, we will have more urban areas extension, more rural areas destructions and more abandoned urban areas because of the new urban areas replace activities located in other older urban areas that will be abandoned generating in this way degraded urban areas.

2.3 The introduction of concept of green infrastructure

Currently it is well known the concept of green infrastructure, consolidated in Europe since it definition by European Commission in 2013 [7]. The Green Infrastructure (GI) is a way to improve natural capital by it integration in territories, including urban and metropolitan areas, such as network of rural and natural lands connected between them. From this point of view, society can improve it sustainability connecting open spaces of urban and metropolitan areas between them.

Also in València this concept is being promoted protecting open spaces that generate ecosystems services and connecting these spaces between them. Some methodologies have been developed to identify and design the GI network in València [8], [9].

Particularly, in València there is a special historic agricultural land, today protected, that today is part of GI system of metropolitan area of València [10], [11]. Also GI is a way to improve sustainability in metropolitan areas where "urban sprawl" make unsustainable forms [12], [13].

Under this theoretical framework, only urban renovation is the way to change the current situation in order to improve the sustainability of urban and metropolitan areas redesigning GI network inside it, or like some authors say, rewilding urban and metropolitan areas.

2.4 Urban touristic developments and the case of Saler's Devesa

In Fig. 2 you can see an image of the touristic urban project designed by Cano Laso architect, according to proposals of Modern Architecture Movement, for urban development of "La Devesa" in 1963.

Figure 2: Cover of journal "Valencia atracción" with an image of touristic urban project, January 1964. *(Source: Municipal Newspaper Archive of Valencia.)*

The urbanization project of "La Devesa del Saler" or Saler's Devesa, was a project to urbanize a coastal area such as way to promote tourism and obtain foreign currency in Spain in the 1960s, in the period of Franco dictator. The case is very interesting because the land was public domain, the whole of public works were executed and paid by city council, a lot of plots was sold to build and a part of them were built. However, in the 1980s, the first democratic municipal government, according to a political program, began the action to return land to city council and rewilding the original ecosystem. It is a case of green infrastructure renovation to maintain a very important ecosystem, its ecological services and, particularly a big natural beach close to Valencia City.

Figure 3: "La Devesa" forest today, with a part already urbanized and built on. *(Source: Google Earth, 05/05/2019 satellite image.)*

The case has been analysed by Martínez Llorens in his Doctoral thesis [1]. It is one of the few cases of rewilding a public forest protected since medieval age and sold to promote a touristic urban area justified on the idea of economic progress. This idea understand social progress such as a way to increasing GDP growth in the short term that implies to increase social and economic activities and jobs without considering negative effects for the future like the destruction of forest ecosystem, and consequently the landscape attraction for foreign tourists and also an high quality natural park next to the beach for local population. In fact, this case shows the change of paradigm about this way of understand the social and economic progress [14].

In contrast with this case there is the case of "La Manga del Mar Menor". This is a similar case also in a sand band between the inland lake of "Mar Menor" and Mediterranean Sea in Murcia department in Spain. This case was analysed by García-Ayllón [15], [17] and Miralles i Garcia and García-Ayllón [16]. It is also a project of touristic political developed in the 1960s in Spain. In this case, like the others touristic projects promoted such as part of official touristic political in this period, was totally executed. The initial project (that was also inspired in proposals of Modern Architectural Movement and elaborated by Antonio Bonet, architect) was progressively transformed on a mass use of residential buildings that generated a very congested urban area without public or private facilities to support touristic activities and also without land to build facilities, infrastructures or equipment. Today only with urban renovation processes would be possible transforming the area to improve quality and facilities, solve environmental problems and maintain touristic activities with a level of quality.

Instead, in the exceptional case of "La Devesa", the majority of public works executed by city council were demolished and were developed several projects to rewilding the majority of area as you can see in Fig. 4 that show the promenade built before demolition and currently landscape with rewilding in process.

(a) (b)

Figure 4: (a) Picture (taken in 1996) of the built promenade; and (b) Picture (taken in 2017) of the process of rewilding after the demolition of the promenade. *(Source: Photographic Archive of Municipal Office "Devesa-Albufera".)*

In addition, it is necessary remember the speculative real estate cycles that took places in Spain over three periods of time: 1960–1972, 1986–1992 and 1997–2007 [18], [19]. These speculative processes were caused mainly by the concept of land propriety to build regulated in Spanish laws. In fact it is contradictory to pay a large quantity of money for a good, the land to build (that is a bad name because the builder or construction company really need the underground to lay the foundation of buildings or public works, and this underground is a natural good not made by anyone) that really is a natural good. The fact of paying for a good that no one made or without no one added value made by anybody, impoverishes society, that is, enriches a few at the cost of impoverishing many that must pay for a product that nobody has really manufactured. In addition, often, people that became rich quickly by this method, instead of by productive economic activity, became an idol for a lot of people in society that want to do the same. In contrast, it has long been known that the only source of wealth is work [20].

3 METHODOLOGY AND DATA

This article shows a part of the results of a PhD research. Particularly about the following secondary objective or question: What should be the idea of society's progress when touristic urban development growth is no longer possible or desirable?

In fact, the case of "La Devesa" is a case where building this space to growth GDP is not possible because of environmental values of the "La Devesa" forest are more important than the economic activities produced through building the area. Valencian society decided by a democratic system rewilding this area. The majority of Valencian society preferred to build other spaces and maintain the ecosystem of "La Devesa". It is very difficult a comparative analysis of economic benefit for society between the two possible situation of "La Devesa" forest: building a touristic city or maintaining it use as natural ecosystem. However, on the other hand, is obviously that, to maintain social and economic activities in urban and metropolitan areas, it is necessary maintain the ecosystems (natural capital) that are producing the environmental services that society use to its own activities. In a general situation, where the expansion of urban and metropolitan areas can produce destruction of natural ecosystems inside them and, at the same time, it is necessary promote and maintain a green infrastructure as a way to sustainability, it is important to know how we can rewilding areas inside urban and metropolitan areas and which can be it economic cost.

In consequence, the research was focused to analyse the rewilding economic cost of the area of "La Devesa" forest.

According to Alía [21], the methodology used is that of scientific method for historical research focused to know the historical process, calculate the cost of building and rewilding and interpret historic acts generating a coherent social logic. In sum, the methodology has been: obtain historical data from different sources about public propriety sold, cost of public works, cost of rebuy plots, cost of demolition and cost of rewilding; semi-structured interviews to significant social agents with protagonist role in actions to identify data sources, obtain original documents, identify social agents and to know the difficulties and the process. You can see more detailed information in Doctoral thesis [1].

The main sources of historical data consulted have been the following:

- Historical bibliography;
- Historical documents about urban activities of city council, especially in Municipal Office Devesa-Albufera, Municipal Office of Urban Planning, Historic Municipal Archive of Valencia and Historic Archive of Valencian Community;
- Urban plans approved or previous studies;
- Rewilding projects;
- Laws applied at all times;
- Land propriety, especially by Public Utility Forest Catalogue, General Inventory of Municipal Properties of Valencia, Land Registry and others;
- Municipal Newspaper Archive of Valencia;

- Brochures of activities or real estate developments;
- Documents of conferences and workshops.

4 RESULTS AND DISCUSSION

4.1 Revenue and costs during building process

At the beginning, land of "La Devesa" forest was propriety of Valencia City Council that obtained permission from "Ministerio de Gobernación" to sell plots for private buildings on 17 February 1970. The sale was undertaken by a public tender system. The first public tender to sell plots for building inside of "La Devesa" forest was done on 10 March 1970 and the last public tender was on 10 January 1973.

Between 1970 and 1973 a total of 410 plots were sold and Valencia City Council obtained an income of 457,345,945 pesetas (approximately 42,997,352 euros currently based on the exchange from 1973 pesetas to 1999 pesetas to 1999 euros to 2020 euros).

The total surface of plots sold was 394,899 m^2 where a total of 1,345,800 m^3 could be built on. Therefore, the average sale price of the right to build was 339.83 pesetas per cubic metre (approximately 31.95 euros per cubic metre) or 1,158.13 pesetas per square metre of plot (approximately 108.88 euros per square metre of plot).

However, only 63 plots out of the 410 was built on, which equates to a total of 103,740 m^2 of plots built according to the following distribution:

- Zone 1. Sidi Saler. Hotel, 1 plot, 200,000 m^2.
- Zone 2. La Casbah. Residential area with single family homes, 36 plots, 12,604 m^2.
- Zone 5. Aparwarks. Apartment towers, 5 plots, 2,718 m^2.

Regarding the public works, 124 certificates of executed public works and 118 certificates of review executed public works have been found. The last certificate was signed in 1978. Consequently we can consider that public works to urbanize the area finished in 1978. The final cost of public works was the following:

- Total public works certificated: 915,023,100 pesetas (approximately 31,456,565.65 euros, currently)
- Total price review of public works certificated: 373,574,248 pesetas (approximately 12,842,695.80 euros, currently)
- Management cost: 2,703,959 pesetas (approximately 92,956,41 euros, currently)
- Total urbanization cost: 1,291,301,307 pesetas (approximately 44,392,217.35 euros, currently)
- Interest for late payment of certifications: 95,786,418 pesetas (approximately 3,292,935.17 euros, currently)
- Total cost urbanization: 2,331,337,725 pesetas (approximately 80,146,477.39 euros, currently)

Considering that the income from selling plots was 1,264,130,620 pesetas, Valencia City Council had a loss of 1,067,207,105 pesetas (approximately 36,688,330.99 euros, currently) as final result of real estate operation.

4.2 Repurchase of plots between 1983 and 2003

The 347 plots sold by Valencia City Council but not build on were repurchased by the first democratic local government. These 347 plots were occupying a surface of 291,159 m^2 while the 63 plots built occupy a surface of 103,740 m^2. Valencia City Council recovered the 347 plots without build throughout period 1983–2000.

Valencia City Council planned to recover the plots at the average price of 3,877.35 pesetas/m^2 calculated as initial sale price plus the increase of the cost of living.

The process of repurchase started in 1983 with the purchase of the first 106 plots by agreement with owners. The rest of the plots were recovered through legal proceedings and court decisions.

Finally in 2003 finished the recovered processes. The final cost of the 347 plots recovered was 2,254,791,846.3 pesetas (approximately 26,967,559.22 euros, currently). That is, the final average price of purchase was 7,744.19 pesetas per square metre (approximately 92.62 euros (currently) per square metre). Fig. 5 shows a detail of location of main plots recovered.

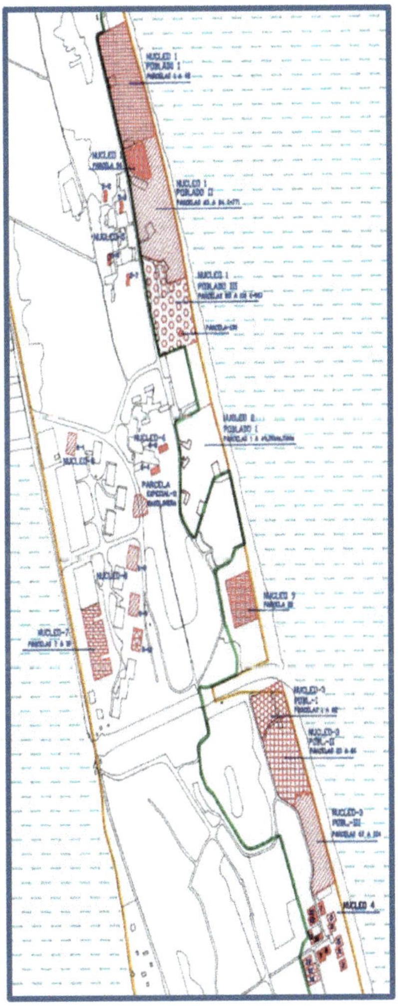

Figure 5: Detailed map with location of plots to recover. *(Source: "Previous report about plots recovery inside 'La Devesa' forest", 2000, Heritage Inventory Office of Valencia City Council.)*

As a consequence, the total surface of segregated plots of initial only plot of "La Devesa" get down to 919,175.99 m^2 distributed according to the following:

- National tourism parador and golf course (723,436 m^2)
- ICONA farm (50,000 m^2)
- Bull-fighting inn (32,000 m^2)
- Stevedores school (10,000 m^2)
- Residential zone 2 "La Casbah" (12,603.99 m^2)
- Residential zone 5 "Aparwarks" (2,718 m^2)
- Residential zone 6 (3,654 m^2)

All of recovered plots had been classified such as Open Spaces System in Valencia Urban Plan approved in 1988. This classification implies that these areas are public domain and cannot be built on.

In addition of public works demolition and plot recovery, Valencia City Council undertook several rewilding works to recovery natural soil and ecosystem. Table 1 shows the rewilding works executed. These natural restoration projects include demolition works of urbanization works.

Table 1: Rewilding areas inside "La Devesa" and cost. *(Source: Vizcaíno et al. [22] and authors.)*

Intervention zone	Surface (m^2)	Cost (euros)
Beach: El Canyar	4,742	87,495.19
Beach: La Malladeta and La Brava	717,012	5,397,123.29
Beach: Els Ferros and La Garrofera	98,675	3,063,190.00
Beach: El Saler	60,000	9,486,330.82
Municipal campsite: El Saler	87,300	ns/nc
Sport area: El Saler	52,000	849,405.00
Wetland: Racó de l´Olla	243,430	1,147,933.12
Total	**1,263,159**	**20,031,477.42**

Adding the debt generated by the urbanization works (6,414,044 euros), the final cost of the recovered plots (26,967,559 euros), the cost of demolition public works of the urbanization and the cost of restoration of habitats in "La Devesa" during the last three decades (20,031,477 euros), the total amount invested in rewilding action amounts to 53.4 million euros, currently. This quantity is an estimate or rough guess at the cost because of the difficulty to do a precise calculation with payments done in different times.

5 CONCLUSIONS AND IMPLICATIONS

The appraisal of "La Devesa" when was sale to the Valencia City Council in 1927 was calculated by capitalizing its annual income in 1911, in fact capitalizing their products at 4%. That is, the value of the land of "La Devesa" before touristic development was associated with the yield of the goods it produced (fan palm, reed, sand reed, esparto grass, low wood, hunting).

This criterion was modified in 1964 by the Law that allowed its urbanization. It responded to a political strategy that consisted of favouring the sale of land for the massive urbanization of coast to promote tourism and obtain foreign currency. In the case of "La Devesa", the Valencia City Council itself (owner of the property) acted as developer of the urbanization works, seeking to offset the expenses derived from them with the income obtained from the sale of public land in the form of building plots.

In other words, the new value of the land was generated by the potential urban development that was attributed to it through the new urban plan. In this way, the value of the land depended exclusively on the price of the buildings that is possible to build (use above the ground). The operation was considered such as driver for economic development due to the jobs it should generate and the possibility of new touristic economic activity. A privileged land close to the city of Valencia was sold to participate in a kind of urban development that was being promoted throughout Spain and that supposedly should contribute to the enrichment of Valencian society. But in fact, the operation was of a characteristic speculative real estate transaction for the benefit above all of a minority of society, to the detriment of the general interest.

But Valencian society faced this criterion, claiming the environmental values of this space in an intuitive and emotional way. The citizen opposition to the urbanization that was destroying this place coincided with the birth of a collective conscience sensitized with the protection of the environment and the start of specific scientific production about "La Devesa", which has remained constant to today. The decision to recover and preserve "La Devesa" was made by the Local Government that emerged from the first democratic elections.

In other words, Valencian society claims the space of "La Devesa" such as a public and natural space of the city of Valencia based on parameters that are not directly economic and difficult to quantify, associated with the improvement of well-being generated by open spaces and protect ecosystems which we now know as green infrastructure. Currently, this environmental awareness of society is well established for the maintenance of the ecosystems of our planet. This idea did not exist at that time, but it has progressively consolidated in the imaginary of our society, increasingly committed to the environment.

The recovery of "La Devesa" meant to reject the economic benefit expected from the urban development and the responsibility of assuming an added cost for the recovery of the soil and its geomorphological restoration.

In fact, this case shows a change of paradigm about how understand the progress of society contrast the economic benefit (usually measurement by GDP) with other intangible benefit such as environmental quality or maintain an specific forest or ecosystem. It is very difficult to quantify these benefits to compare with economic benefits.

On the other hand, this case shows the difficulties of rewilding processes or urban renovation processes. In fact, in this case it was not possible recovery the built plots because it was too expensive. This fact focus the attention on the role of private propriety of land to build. In a situation which metropolitan areas are in expansion and there is the need of urban renovation to improve urban sustainability and develop green infrastructures, recovery urban plot to change it use is, currently, very expensive when, in fact, builders need the underground, that is a natural good, to build. But not only this, other problem especially important is the landowners' right when the building is demolished. Usually new builders or administration have to pay a lot of money for the rights to build associated to underground. In this conditions it is very difficult urban renovation processes. In fact, now in Spain, urbanize rural areas is cheaper than renovation old urban areas, although old urban areas have all kind of urban services.

This historical case contrast with also historical case of "La Manga" and allows to compare both models. In fact, "La Manga" model present a list of problems of degradation with very difficult and expensive solutions by urban renovation processes. Therefore, if the economic analysis of urban tourist project in a specific location is only a direct analysis cost–benefit without consider rewilding cost of previous ecosystem services, really the measuring of progress by immediate increase of GDP is false because of the real cost has been transferred to the future.

This original research is focused in a rewilding or urban renovation case. The authors doesn't know other published researches about similar cases with analysis about economic costs. Therefore this research open new research lines:

- How can we calculate or measure the society progress including cost of maintaining ecosystems services necessaries to own social and economic activities?
- Can the cost of rewilding have a tool to include the real cost of projects that destroy naturals ecosystems and it ecosystem services?
- Is it possible to identify ecosystems that produces environmental services and to calculate it value as cost of rewilding after it destruction?
- Is it possible to identify a natural capital and its value in this way?

ACKNOWLEDGEMENTS

The authors wish to express their gratitude to the Municipal Devesa-Albufera Service, the Municipal Heritage Service and the Municipal Planning Service of Valencia city for their assistance in facilitating the consultation of the historical documents, which enabled the execution of the doctoral thesis, the origin of this paper.

REFERENCES

[1] Martínez Llorens, F., Devesa de l'Albufera: El cambio de paradigma en el turismo de masas de la ciudad de Valencia. PhD thesis, Universitat Politècnica de València, 2019. https://riunet.upv.es/handle/10251/121137. Accessed on: 6 May 2020.

[2] Vheeler, S.M., The evolution of built landscapes in metropolitan regions. *Journal of Planning Education and Research*, **27**, pp. 400–416, 2008. www.//jpe.sagepub.com/cgi/content/abstract/27/4/400. Accessed on: 20 Feb. 2017.

[3] Aguilera, F., Valenzuela, L.M. & Botequilha-Leitão, A., Landscape metrics in the analysis of urban land use patterns: A case study in a Spanish metropolitan area. *Landscape and Urban Planning*, **99**, pp. 226–238, 2011. www.sciencedirect.com/science/article/pii/S0169204610002823. Accessed on: 20 Feb. 2017.

[4] Feria Toribio, J.M. & Martínez Bernabeu, L., La definición y delimitación del sistema metropolitano español: permanencias y cambios entre 2001 y 2011. *Revista Ciudad y Territorio. Estudios Territoriales*, **187**, pp. 9–24, 2016.

[5] Miralles i Garcia, J.L., Environmental and territorial planning on coastal metropolitan areas. A methodological proposal for València (Spain). *International Journal of Environmental Impacts*, **2**(1), pp. 1–16, 2019.

[6] Hamilton, S., *Cultivating Nature: The Conservation of a Valencian Working Landscape*, University of Washington Press: Seattle, 2018.

[7] European Commission, Communication from the Commission to the European Parliament, the Council, the European Economic and Social Committee and the Committee of the Regions. Green Infrastructure (GI): Enhancing Europe's Natural Capital. http://eur-lex.europa.eu/legal-content/EN/TXT/?uri=CELEX:52013DC0249. Accessed on: 14 Oct. 2019.

[8] Miralles i Garcia, J.L. & Altur Grau, V.J., Updated method of aptitude to sustainable urban development for including green infrastructure. *International Journal of Sustainable Development and Planning*, **11**, pp. 970–979, 2016.

[9] Miralles i Garcia, J.L., Green infrastructure on the Mediterranean Valencian coast. *International Journal of Sustainable Development and Planning*, **11**(3), pp. 227–235, 2016.

[10] Miralles i Garcia, J.L., Environmental management of peri-urban natural resources: L'Horta de València case study. *WIT Transactions on Ecology and the Environment*, vol. 192, WIT Press: Southampton and Boston, pp. 99–110, 2015.

[11] Temes, R. & Moya, A., Typology of the transformations occurred in the peri-urban space of Huerta de Valencia. Evidence from North arch of Valencia. *International Journal of Sustainable Development and Planning*, **11**(6), pp. 996–1003, 2016.

[12] Gielen, E., Riutort-Mayol, G., Palencia-Jiménez, J.S. & Cantarino, I., An urban sprawl index based on multivariate and Bayesian factor analysis with application at the municipality level in València. *Environment and Planning B: Urban Analytics and City Science*, **45**(5), pp. 888–914, 2018.

[13] Gielen, E., *Costes del "Urban Sprawl" para la administración local*, Publicacions de la Universitat de València: València, 2016.

[14] Miralles i Garcia, J.L. & Martínez Llorens, F., Tourist development and planning on Valencian Mediterranean coast: The case of "La Devesa del Saler". *WIT Transactions on Ecology and the Environment*, vol. 217, WIT Press: Southampton and Boston, pp. 495–507, 2018.

[15] García-Ayllón, S., Retrospective analysis of urban development In the Spanish Mediterranean Coast. WIT Transactions on Ecology and the Environment, 179, pp. 291–302, 2013.

[16] Miralles i Garcia, J.L. & García-Ayllón, S., The economic sustainability in urban planning: The case of La Manga. *WIT Transactions on Ecology and the Environment*, vol. 173, WIT Press: Southampton and Boston, pp. 279–290, 2013.

[17] García-Ayllón, S., La Manga case study: Consequences from short-term urban planning in a tourism mass destiny of the Spanish Mediterranean coast. *Cities*, **43**, pp. 141–151, 2015

[18] Miralles i Garcia, J.L., Real estate crisis and sustainability in Spain. *WIT Transactions on Ecology and the Environment*, vol. 150, WIT Press: Southampton and Boston, pp. 123–132, 2011.

[19] Miralles i Garcia, J.L., *El darrer cicle immobiliari al País Valencià. O el progrés de la misèria*, Fundació Nexe: València, 2014.

[20] Smith, A., *An Inquiry into the Nature and Causes of the Wealth of Nations*, Vol. 1, 1776. https://books.google.es/books?id=C5dNAAAAcAAJ&pg=PP7&redir_esc=y#v=onepage&q&f=true. Accessed on: 28 May 2020.

[21] Alía, F., *Métodos de investigación histórica*, Síntesis: Madrid, 2016.

[22] Vizcaíno, A. et al., *La Devesa de l'Albufera de Valencia: un caso de restauración dunar. Restauración y gestión de sistemas dunares. Estudio de casos*, ed. F.X. Roig-Munar, Càtedra d'Ecosistemes Litorals Mediterranis, pp. 115–136, 2016.

TOURISM CARRYING CAPACITY FOR BEACHES IN TULUM NATIONAL PARK, MEXICO

MARÍA DENY D POOT QUINTAL & ROMANO G. SEGRADO PAVÓN
Universidad de Quintana Roo, Mexico

ABSTRACT

The tourist use of natural spaces causes multiple, simultaneous, short and long term impacts on the environment, which is why it is necessary to establish limits of use, as in other social activities. The objective of the study was to establish the tourism carrying capacity of two beaches of Tulum National Park (TNP), which have ample tourist and recreational visitation. The multi-criteria method was carried out with a SWOT, satellite images, and interviews with key informants. As a result, it was determined that the maximum limit was 976 simultaneous users for the two most visited beaches of the TNP, Santa Fe and Pescadores, with four use zones with different densities. The public institutions will be able to design better strategies for the conservation of natural spaces and achieve greater satisfaction of the users of these natural spaces.
Keywords: tourism, carrying capacity, beaches, Tulum.

1 INTRODUCTION

The promotion of tourism is mainly based on the attractiveness of its natural resources and landscapes, built in sensitive ecosystems, which consequently can influence degradation processes in reefs, mangroves, forests, beaches, and any pristine site. In this sense, unplanned, chronic, intensive and extensive tourist visitation negatively affects natural spaces, such as the beaches of the Tulum National Park (TNP).

At the international level, tourist visitation to national parks has become a mercantilized process, which under the premise of economic and social benefits, facilitates the use of the landscape as a commodity that creates spaces of social segregation, disarticulated and differentiated in goods and services, which increases levels of population marginalization and promotes forms of consumption of the territory that only favor those who have more [1]. In addition, national parks face increasing pressure to use contiguous zones of influence for the construction and development of residential areas [2]. TCC studies on tourist beaches are limited and almost exclusively based on social perception criteria, omitting zoning and ecological factors.

The commercialization of natural spaces, by criteria of landscape, recreation, social trends, among others, has evolved in such a way that it exerts chronic negative impacts on the natural environment in natural spaces. In the case of Tulum National Park, the tourist activity promoted under the slogan of nature contemplation has generated a series of problems such as the increase in hotel infrastructure, perception of overcrowding among visitors, changes in sea turtle behavior, affecting coastal dunes, as well as multiple situations of illegal use of the territory.

Tulum National Park received 1,773,929 visitors in 2016 [3], who visited the archaeological zone inside the Park and adjacent to the beach area, where 18 hotels, cabins or villas are also present, representing 420 rooms [1]. In this same space, the fishing and tourism cooperatives carry out activities, which are carried out by the inhabitants. Other spaces are also found, integrated by restaurants, beach clubs and other companies of tourist services of support. The immediate zone of influence located to the south of the Park, with a strip of approximately 7 km of coastline; it is complemented with fishing activities; small

WIT Transactions on Ecology and the Environment, Vol 248, © 2020 WIT Press
www.witpress.com, ISSN 1743-3541 (on-line)
doi:10.2495/ST200161

hotels prevail. Along this coastal strip there are 46 hotels with 546 rooms, as well as houses conditioned for lodging service [1].

The TNP constitutes the only terrestrial Natural Protected Area (NPA) in the main tourist corridor of Cancun – Tulum (Riviera Maya). The main tourist attraction is the archaeological zone of Tulum and the nearby beaches: Santa Fe, Pescadores, Maya. While the archeological zone has access control and regulations for visitation, on the beaches there is no registration or access control for visitors and environmental regulations are minimal or non-existent. This situation promotes several situations that negatively impact the TNP.

Establishing the Tourism Carrying Capacity (TCC) for the conservation of natural spaces with visitation is a complex problem because the boundaries are multidimensional, contiguous, non-linear, polygens and heterogeneous, which perform various functions of rights, territorialization and regulation in social and ecological communities. In the case of National Parks, limits are a regulatory strategy observed in the form of norms, rules and principles, created to regulate, prevent and control interactions between tourists and their environment, establishing a standard for social behavior in a given area, in the ecological environment and social dimensions [4].

The World Tourism Organization [5] argues that in order to avoid or mitigate negative impacts it is necessary to establish the carrying capacity, to determine the maximum use that can be made of a natural space without damaging its resources, to decrease the levels of satisfaction of tourists or to generate negative impacts on society, the local economy or culture. With this foundation and in order to operationalize the paradigm of sustainable development and maintain the ecological capacities of resilience, resistance and recreation, the following research question was proposed: what is the tourism carrying capacity of the beaches of Tulum National Park? Developing tourism management strategies helps to avoid translimitation, mitigate negative effects and achieve socio-ecological stability of natural spaces.

2 THEORETICAL FRAMEWORK

The fundamental social conduct consists of not harming the interests of others, either by express legal provision or tacit understanding, and assuming part of the work and sacrifices to be made to protect the social interest. On this basis, when a person or some social group affects the interests of others, society can judge the rights constituted over a territory and the question of imposing or interfering with the limits to the general welfare is open to discussion in Mill [6].

According to Ostrom and Merino [7] society must create and enforce social norms, because there is a retribution, besides being indispensable for social continuity, through stable institutions that establish understandable limits, clear norms and supervision, for the sustainable use of natural resources. This argument, contrary to the one put forward by Hardin [8] in his classic "Tragedy of the commons", establishes that an organized social group can achieve a measured use of natural resources, considering balanced social and individual proportions.

The creation of social norms and limits seek to achieve a balance between the social and individual benefit of nature, by means of sanctions, laws, norms and other forms of behavior regulations, within certain ranges of acceptability, to stimulate cooperation and strategies that lead to communal welfare through the maintenance of the ecosystem [7].

Boundaries present themselves in multiple forms and can be influenced by administrative, social, ecological, cultural, even historical reasons [9], [10]. In human intervention on nature, adaptive management involves the determination of "adaptive limits", which is defined as "the point at which an actor's objectives cannot be assured of intolerable risks through adaptation actions" [9], therefore, the limit was reached and there is no certainty of maintaining the ecosystem service in its outstanding conditions [10].

In the case of tourism, it involves limiting the access of tourists to certain natural areas, for reasons of public interest, either by strict or flexible rules, such as tourism carrying capacity and limits of acceptable change, management techniques and planning, In natural areas, zoning, limits of acceptable change and tourism carrying capacity, are applied to estimate the limits of resistance and resilience for fauna and flora with public use.

In the state of the knowledge, diverse authors have analyzed the tourism carrying capacity and the limits of acceptable change, with different methods. The analysis of the scientific literature revealed that the main paradigm applied was sustainable development [11]–[13].

As a research method, the elaboration of thematic maps for tourist routes and dispersion as a way of diversifying the tourist space and its use has a broad development with territorial emphasis [11], [12], [14]. On the other hand, from social analysis, visitor satisfaction and the perception of overcrowding have been the most prominent factors in studies with surveys and indicators [15]–[17]. In general, it can be stated that the trend in the TCC and LAC studies is multidisciplinary, for a better understanding and application of the variables [13].

In relation to the problems tackled, five relevant aspects stand out: (1) lack of consideration of natural limitations [11], [13], [16]; (2) visitors' influence on the affectation of nature [11]–[14]; (3) concentration of tourist activities [12], [15]; (4) economic relevance of tourist activity [18]; and (5) visitor perception and satisfaction level [11], [15]–[17], [19].

Among the main variables identified are the level of overcrowding [11], [15]; level of satisfaction [17], [19] and number of visitors [12]–[14], [16], [18]. Territory or physical space is the common factor in virtually all studies.

In general, the models for TCC and LAC determination reflect porosity and interaction in determining limits that are considered adequate, either by ecological or social criteria, as a theoretical contribution to sustainability. However, because variables of primary interest cannot be measured directly, data on restrictive factors are used and aspects of community governance must be properly developed to promote conservation.

3 METHOD

Tulum National Park is located in Quintana Roo, in the municipality of Tulum, has an area of 664 hectares, which are distributed mainly along the coastal area, almost in parallel with federal highway 307. It is the only NPA between Cancun and Tulum and is located around the archaeological zone of Tulum [20]. Inside the Park there are three public beaches: Santa Fe, Pescadores and Playa Maya. The first two were selected for this research, due to their wide visitation, both by tourists and the local population. The extension of both beaches is approximately 630 m and the crystalline waters with a turquoise blue color, as well as the fine sand, soft and white color, with palms.

The descriptive method was applied, because territorial data were measured, collected and evaluated, as well as determining the number of people that can support the study area without suffering permanent or short-term damage to its natural environment. Because this study is subject to natural seasons such as summer and winter, it is considered to be transversal.

The focus of this research is quantitative, because for the achievement of the objectives, a serie of field observations are carried out that measure the space covered by the two public beaches inside the Park, in Santa Fe and Pescadores; which allows a numerical analysis. The study was based on the method established by Cifuentes Arias et al. [21] which is widely documented and available on the Internet, so no further descriptions are presented in this regard. This method was complemented with Betancourt and Herrera-Moreno [22], who established a zoning by uses of the coastal region (beach area).

The territory is the basis of the analysis for the physical carrying capacity, after which a zoning by uses of the coastal region of study is carried out. Betancourt and Herrera-Moreno [22] affirm that the subdivision of the beach zone (Fig. 1) into areas allows its management through the assignment of occupation criteria, use and functional relations between the zones that conform it to achieve an optimal use of the beach resource based on a great variety of criteria that are observed in Table 1.

In Fig. 2, the zones established within the area of the beaches to be studied.

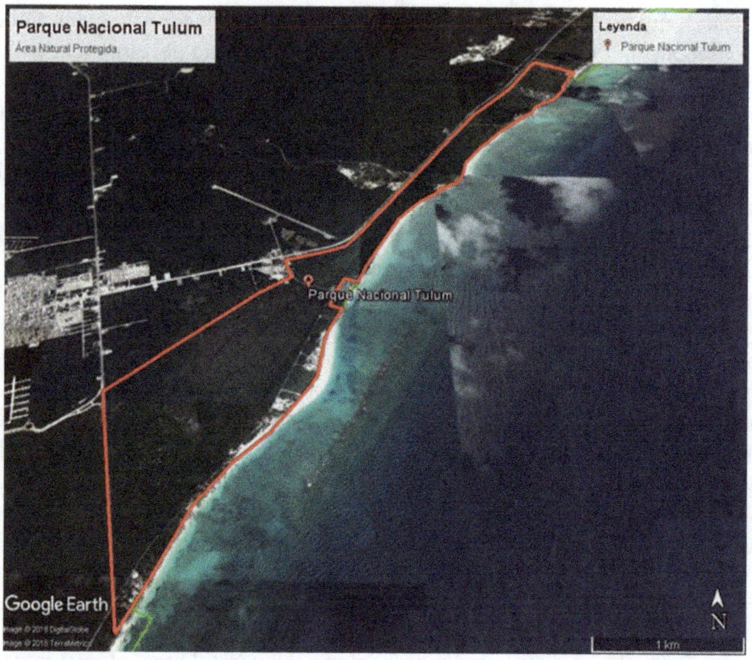

Figure 1: Limits of Tulum National Park. (*Source: own elaboration based on Google Earth, 2018.*)

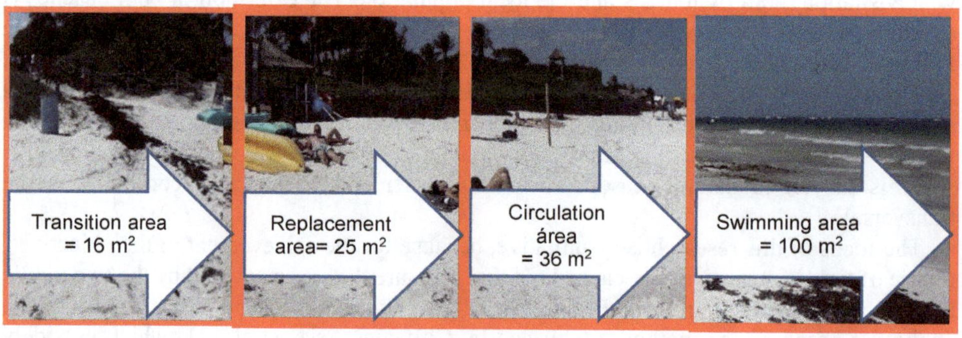

Figure 2: Zoning of the beach. *(Source: Betancourt and Herrera-Moreno [22].)*

Table 1: Table Zoning criteria by uses of the coastal zone. (*Source: Betancourt and Herrera-Moreno [22].*)

Zone	Characteristics	Activities
Swimming area	Area where the visitors, alone or in groups, are located in a quiet way. It is an area of 100 m^2 per user for a depth of less than 1.6 m.	Visitors can sit or stand for a bath or chat.
Intensive use or circulation	Strip of beach where the tides intervene, and that are in constant contact with the sea and that possesses firmness to the step. Its width depends on the slope of the beach and the amplitude of the tide. It is an area of 36 m^2 per user.	In and out of the swimmers, walks among the foam of the waves and where different types of games are practiced.
Resting or sun exposure	A strip of sand above the high tide limit, where the geomorphological characteristics of the beach vary. Its length and width determine the maximum physical carrying capacity of the beach. It is an area of 25 m^2 per user.	The users rest and sunbathe taking advantage of the existing vegetation or using installed infrastructures and recreational, sports or rest activities are carried out that do not need permanent facilities.
Transition to the beach	Strip far from the coastline, behind the resting area. Its transverse extension varies from the area of solar exposure. It is an area of 16 m^2 per user.	Entry, exit and search for a space in the beach area, play and rest activities are carried out, in the sun or in the shade, and service centers can be installed. The activities are carried out informally, although support facilities can be found.

According to Betancourt and Herrera-Moreno [22] the areas where users sunbathe and bathe are not excluded, but are part of the same physical unit and therefore must be intertwined with each other, so that it does not happen that visitors who enjoy the sun do not find space within the bathing area or vice versa.

The eqn to determine the Physical Capacity in each zone of the beach is the following: PCC (Number of users in the beach area) = Surface area of the area (m^2)/Occupancy criterion (m^2/user).

For the calculation of this area, the modified occupancy criterion established by Mexican Standard NMX-AA-120-SCFI-2016 [23] was adopted, which defines the beach as a geomorphological unit formed by the accumulation of unconsolidated sediments of different types and whose limits are established in lower and upper limits. Once the PCC has been established and the areas of use on the beach have been determined, the actual carrying capacity of the study site can be evaluated.

The measurement of correction factors, or "critical limiting factors" is of great importance for the calculation of the actual Carrying Capacity. The correction factors according to Segrado and Arroyo [24] are certain restrictions or limitations to the CC directly linked to some aspect of the tourist attraction, be it physical, ecological, management, safety, visitor satisfaction.

The natural correction factors identified for this research were the following elements: sea turtle nesting seasons; yellow or red flags; algae and red tides; sargasso, because it diminishes the enjoyment space for visitors. It is worth mentioning that these factors are subject to annual temporality and are not permanent. On the other hand, the identified social factors presented very subjective characteristics, so all were discarded, with the exception of the minimum personal space for the enjoyment of the beach, in which a certain consensus was achieved.

With respect to the management capacity, the restrictive aspects, but not limiting, were the capacity for emergency attention (personnel and schedule); sanitary (optimal density), water treatment; freshwater showers; awning service; educational induction program; program for monitoring environmental variables and administration program, palm shelters and changing rooms.

4 RESULTS

The physical carrying capacity was determined by the available surface area, the area occupied by one person and the number of times the site can be visited by the same person in one day. In this regard, it was considered appropriate to select the minimum distance of 16 m^2 per person, a daily schedule of attention to visitors in the study area was carried out from 7 a.m. to 7 p.m. making a total of 12 hours of attention, likewise it was determined that the average time of visit is 8 hours. With the substitution of the previous data, NV = 12 hours day / 8 hours average visit = 1.5 visit per day per visitor.

Once the data were collected, the substitution of values was applied to determine the Physical Carrying Capacity (PCC): PCC = 82,269.207 m^2 x 1.5 probable visit per day / 16 m^2 per person = 7,712 people per day. However, this preliminary figure must be analyzed considering the zoning of the study area according to their respective measures (see Table 2).

The division of the beach in areas with different types of uses, allows its arrangement by establishing occupation criteria and functional relations between the areas that structure it to achieve optimal use of the beach resource. To this end, four zones were established. Therefore, by applying the zoning criteria, and then adding them together, the PCC of the beach is obtained. The eqn for zoning is as follows: Area (m^2) / Occupancy criterion (m^2/user) = Number of users in the beach area.

The data considered in the zone of "transition to the beach" were replaced, with a total of 30,299.117 m^2 and a criterion of 16 m^2 per person, the eqn was applied and obtained as a result: 1,893 people. Similarly, the data considered in the area of "rest or sun exposure", with a total of 13,134.773 m^2 and a criterion of 25 m^2 per person, the eqn was applied and obtained as a result: 525 people. The data considered in the area of "intensive use or circulation" were substituted, with a total of 13,340.376 m^2 and a criterion of 36 m^2 per person, and the result was: 370 people. Finally, the eqn was applied with the data considered in the area of "swimming area", with a total of 25,494.941 m^2 and a criterion of 81 m^2 per person, obtaining as a result: 314 people. Adding the subtotals for each area (1,893 people + 525 people + 370 people + 314 people) gives the PCC: 3,102 people. Table 3 summarizes the zoning results.

In order to obtain the Real Carrying Capacity (RCC), it was necessary to establish the natural and social Correction Factors (CF). Among the former were nesting seasons (turtles), flags, algae, red tide, and serge; among the latter were garbage bins, tables, chairs, beds, umbrellas, palapas, lifeguard tower, and boats. To calculate the total area of enjoyment space, the total available area of 82,269.21 m^2 – 4,754.48 m^2 of the total area limiting the correction factors was considered, resulting in 77,514.73 m^2 of enjoyment space, which represents 5.78% of the total available area of the zones.

The beaches are subject to seasonal and interannual variations of natural processes such as the case of the identified natural correction factors: flags, algae, red tide and sargasso. These natural phenomena begin spontaneously and in an unpredictable way, and have a variable duration, although in some sectors they occur with a certain periodicity and in others occasionally; it happens mainly in summer seasons although in the last four years its periodicity has increased, with the consequence that the beaches of these states are temporarily closed, which is why it is an important factor to be considered in this study, as it prevents immersion in water, even in areas close to the beach. From the correction factor "waves", during seasons of strong winds there is a change in the frequency and strength of tides, generating that the access to the sea is limited or better known as bathing area. As a preventive measure, visitors are warned by the use of flags: red (danger), yellow (precaution) and green (free use).

Once the factors are determined, the values are substituted to determine the actual carrying capacity (RCC), whose eqn is RCC = [(PCC x (100% – CF)]. So the RCC = ((3,042 people per day x (100% – 5.78%)) = (3,102 x 94.22%) = 2,922 people per day. Thus, the actual Carrying Capacity of the area is 2,922 people per day for the study area.

In the study area, which includes the public beaches of Santa Fe and Pescadores, there are restaurant services; bathrooms (private), lifeguard tower, showers (private), hotel, beach club (includes hotels and cabins that call themselves beach clubs, due to the extension of the beach on which they have a concession).

Table 2: Study area zoning. *(Source: Based on Qgis 3 and Google Earth Pro, 2019.)*

Zoning	**Area in m^2**
Transition area	30,299.117
Rest area or sun exposure	13,134.773
Area of intensive use or circulation	13,340.376
Swimming area	25,494.941
Total	82,269.207

Table 3: Beach zoning (*Source: Prepared by the authors, 2019.*)

Zoning	**Criteria**	**Total people**
Transition area	16 m^2 per person	1,893
Rest area or sun exposure	25 m^2 per person	525
Area of intensive use or circulation	36 m^2 per person	370
Swimming area	81 m^2 per person	314
Total		3,102

It was also considered the optimal infrastructure for the administration to offer a quality service for visitors and of which the beaches that were part of the research do not count: parking, dressing rooms, viewpoint, access for people with disabilities, palm shelters (to protect from the rays of the sun), kiosks for rent of umbrellas, chairs and tables, among other factors. The percentage value, as a comparison between the real and the ideal, was 29.95%.

The equipment used to carry out the activities of the personnel working in the study area includes the working tools available to the lifeguards: first aid kit, rescue buoys, torpedoes, whistles, radios, ATVs; while the police present buggy, armament and radio vehicles. The percentage value of the comparison of the ideal equipment against the real one is 50.35%.

People hired to support or provide services to visitors collaborate with the quality of visitation. In the public beaches of Santa Fe and Pescadores there are two guards, two police officers and two park rangers; although due to the increase in the recharge of serge, the need for cleaners was identified for this affectation. The percentage value of the real towards the ideal reached 12.50%.

The administration of the public beaches of Santa Fe and Pescadores has not established any type of cultural or educational activities, although occasionally garbage collection and turtle release days are held, known as environmental education activities. The percentage value for the qualitative value between the ideal and the real reaches 41.67%.

To obtain the total value for the Administration category, the average of all the resulting factors is obtained and the eqn is applied. Substituting: Adm. = 29.95% + 50.35% + 12.50% + 41.67% x 100 / 4 = 33.62%. Therefore, the management capacity of the administration is below that established for the achievement of adequate enjoyment of the recreational site by visitors.

With regard to food, the total capacity is 312 chairs and food production allows to feed this number of people every 60 minutes, so the Food indicator is equal to 100%. There are only two private bathrooms in the study area, specifically on the public beach of Santa Fe, but there are no data on septic tanks, so the value of 0% was assigned to this criterion.

In the public beaches of Santa Fe and Pescadores there are a total of 12 bins for the deposit of garbage in general, without indicating the type of waste or residue that can be deposited in them. There is no program or containers for the separation of solid waste or waste recovery, therefore the value for this criterion is 0%.

Once the values of the indicators that make up the Management Capacity (MC) were obtained, the eqn was applied: 33.62% +100% + 0% + 0% x 100 / 4 = 33.40%. Therefore, the management of activities and achievement of objectives and goals of the administration of the public beaches Santa Fe and Pescadores, have limitations that affect the RCC.

The Effective Carrying Capacity (ECC) is the maximum limit of visits that can be allowed, based on the resources of the local administration to order and manage them. The eqn CCE= 2,922 x 33.40 / 100 was applied, resulting in 976 daily visitors to the public beaches of Santa Fe and Pescadores. This figure is the maximum daily number of people who can make use of the tourist site under the current conditions analyzed.

5 DISCUSSION

It is a complex process to establish the number of visitors that an area can support without being affected, because there is a lack of consideration of natural limitations and concentration of tourist activities in a small number of places, giving greater relevance to economic development that encourages tourism than to conservation and protection of the protected natural area. The variety of parameters to be considered when determining the carrying capacity is another of the difficulties that must be faced in carrying out the research;

among which are: type of activity, seasonality, schedule, conservation status of the natural area's resources, existing facilities and installations, degree of user satisfaction, and so on.

The studies by Lobo et al. [18]; Collins-Kreiner et al. [11]; Malik and Bhat [12] and Lobo [14] are those that agree with this research since the purpose is to establish the carrying capacity as a method to be applied to determine the spatial correlation and distribution of the tourist space; based on an analysis and the establishment of an adequate tourist planning of the natural areas that allow the diversification of the tourist activity in Tulum National Park. However, studies by Rathnayake [15]; Giglio et al. [16] and Zhang et al. [17] helped to complement the carrying capacity in a multidisciplinary manner to have a better understanding and adaptability of the variables [13]. However, aspects such as satisfaction levels and visitor perception were discarded, since for this research determining the level of satisfaction or perception was not one of the established objectives.

Betancourt and Herrera-Moreno [22] zoned beaches with three criteria, which are: intensive use or circulation, rest or solar exposure and solar transition. However, for this research it was considered pertinent to add the bathing area, because it is a real factor of the physical capacity of the beach, since at some point during their stay, the visitor may be inside the sea, expanding the space and therefore increasing the FCC.

The main finding is that there is a disproportionate increase in visitor service infrastructure in the area and its area of influence, in contrast to its management capacity of the administration, which remarked to 33%. What is surprising is that there is not a management plan, despite being a protected natural area with the designation of National Park, whose priority should be the protection of its natural resources, the increase of its flora and fauna, and in general, the preservation of ecosystems and their elements [25]. It was also identified that natural correction factors such as sargasso are currently limiting the carrying capacity of the study area, despite the fact that social factors play a relevant role.

6 CONCLUSION

This research determined the tourism carrying capacity of Tulum National Park, specifically in its primary zone of influence, which includes the public beaches of Santa Fe and Pescadores. The result is 976 visitors as the maximum limit of visits that can be made to the site during each day.

In order to carry out this research, three specific objectives were proposed; among which is the establishment of indicators for Tourism and Recreational Carrying Capacity. For this reason, the study area was zoned in four zones, which are: transition area, rest or solar exposure, intensive use or circulation and bathing area, from which the correction factors were determined both natural and social of each. Therefore, in the transition area, garbage cans, tables, chairs, beds, umbrellas, palm shelters, and lifeguard tower were identified as limiting areas; likewise, the nests of turtles and motorboats were identified as limiting areas for resting; likewise, in the bathing area, only motorboats were identified as limiting areas; as for the area of intensive use, limitations were not identified due to the fact that it is imperative to leave it free for visitors to move from one area to another.

Another specific objective was to determine the natural and social factors that determine the Tourism Carrying Capacity of Tulum National Park. As for the social correction factor identified within the park, there are garbage cans, tables, chairs, beds, umbrellas, palm shelters, lifeguard towers, and boats since these elements occupy spaces that limit visitors' areas of enjoyment. The natural correction factors were the nesting season, flags, sargasso, algae and red tide; since they reflect a limitation of enjoyment space and also limits the activities that tourists can develop, it is worth mentioning that everything described above is subject to annual temporality and does not occur all year round.

Determining the Tourism Carrying Capacity of the public beaches of Santa Fe and Pescadores will help establish limits on tourism development in the area, which in turn will contribute to better planning and tourism management by managers based on the carrying capacity of the ecosystems, thus facilitating decision making and the design of strategies to manage the tourist use of the TNP. Because the research contributes concretely to Tulum National Park as a prevention tool that contribute to better tourism development, achieving true sustainability of the study area allowing a better distribution of economic benefits among communities and sectors involved.

ACKNOWLEDGEMENTS

Special thanks to the National Council of Science and Technology (CONACYT), to the University of Quintana Roo, Cozumel Unit; for the support, the means and facilities granted to obtain the degree and development of this research work.

REFERENCES

[1] Arroyo, L., Segrado, R. & López, Á., Estructura hotelera como modificadora del espacio litoral en Tulum, México, 2019. www.redalyc.org/articulo.oa?id =456144904007. Accessed on: 12 Nov. 2019.

[2] Arnberger, A. & Brandenburg, C., Past on-site experience, crowding perceptions, and use displacement of visitor groups to a peri-urban national park. *Environmental Management*, **40**(1), pp. 34–45, 2007.

[3] SEDETUR Indicadores Turísticos Enero – Diciembre 2016. Quintana Roo 2017. http://sedetur.qroo.gob.mx/estadisticas/indicadores/2017/Indicadores%20Tur%20-%20Enero%20-%20Diciembre%202016.pdf. Accessed on: 6 Nov. 2017.

[4] Segrado, P.R. & Chuc Piña, I., Sustainable tourism development in Parque Nacional Lagunas de Montebello, México. *WIT Transactions on Ecology and the Environment*, vol. 8, pp. 83–93, 2018. www.witpress.com/Secure/elibrary/papers/ST18/ST18009FU1.pdf

[5] Indicadores de desarrollo sostenible para los destinos turísticos, Madrid, España: WTO elibrary, 2005. www.ucipfg.com/Repositorio/MGTS/MGTS14/MGTSV-07/tema2/OMTIndicadores_de_desarrollo_de_turismo_sostenible_para_los_destinos_turisticos.pdf. Accessed on: 25 Feb. 2019.

[6] Mill, J., *On Liberty*, Longman, Roberts & Green: London, 1869.

[7] Ostrom, E. & Merino, S.L., *El gobierno de los bienes comunes*, México: Fondo de Cultura Económica, 2011.

[8] Hardin, G., The Tragedy of the Commons The population problem has no technical solution; it requires a fundamental extension in morality, Science, 1968. https://science.sciencemag.org/content/sci/162/3859/1243.full.pdf. Accessed on: 18 Feb. 2019.

[9] Dow, K., Berkhout, F., Preston, B.L. & Klein, R.J.T., Limits to adaptation. *Nature Climate Change*, **3**(4), pp. 305–7, 2013.

[10] Klein, H.J., Cooper, J.T., Molloy, J.C. & Swanson, J.A., The Assessment of commitment: Advantages of a unidimensional, target-free approach. *Journal of Applied Psychology*, **99**(2), pp. 222–238, 2014.

[11] Collins-Kreiner, N., Malkinson, D., Labinger, Z. & Shtainvarz, R., Are birders good for birds? Bird conservation through tourism management in the Hula Valley, Israel. *Tourism Management*, **38**, pp. 31–42, 2013.

[12] Malik, M.I. & Bhat, M.S., Sustainability of tourism development in Kashmir – Is paradise lost? *Tourism Management Perspectives*, **16**, pp. 11–21, 2015.

[13] Maia, T. & Smith, C., Scenario-based planning for tourism development using system dynamic. *Tourism Management*, **68**, pp. 336–354, 2018.
[14] Lobo, H.A.S., Tourist carrying capacity of Santana cave (PETAR-SP, Brazil): A new method based on a critical atmospheric parameter. *Tourism Management Perspectives*, **16**, pp. 67–75, 2015.
[15] Rathnayake, R.M., How does 'crowding' affect visitor satisfaction at the Horton Plains National Park in Sri Lanka? *Tourism Management Perspectives*, **16**, pp. 129–138, 2015.
[16] Giglio, V.J., Luiz, O.J. & Schiavetti, A., Marine life preferences and perceptions among recreational divers in Brazilian coral reefs. *Tourism Management*, **51**, pp. 49–57, 2015.
[17] Zhang, Y., Li, X.R. & Su, Q., Does spatial layout matter to theme park tourism carrying capacity? *Tourism Management*, **61**, pp. 82–95, 2017.
[18] Lobo, H.A.S. et al., Projection of tourist scenarios onto fragility maps: Framework for determination of provisional tourist carrying capacity in a Brazilian show cave. *Tourism Management*, **35**, pp. 234–243, 2013.
[19] Zhang, Y., Li, X.R., Su, Q. & Hu, X., Exploring a theme park's tourism carrying capacity: A demand-side analysis. *Tourism Management*, **59**, pp. 564–578, 2017.
[20] Tulum a la Vanguardia. Parque Nacional Tulum. Una muralla abierta a la conservación, 2007. www.conanp.gob.mx/conanp/dominios/tulum/. Accessed on: 8 Nov. 2017.
[21] Cifuentes Arias, M. et al., Capacidad de Carga Turística de las Áreas de Uso Público del Monumento Nacional Guayabo, Costa Rica. WWW Centroamérica, 1999. http://awsassets.panda.org/downloads/wwfca_guayabo.pdf. Accessed on: 23 Nov. 2017.
[22] Betancourt, F.L. & Herrera-Moreno, A., Acercamiento de la Capacidad de Carga Física de Playa Grande, Cayo Levantado, Samaná, República Dominicana. Programa EcoMar República Dominicana, 2005. https://programaecomar.webs.com/CayoLevantado.pdf. Accessed on: 23 Nov. 2017.
[23] Norma Mexicana NMX-AA- 120-SCFI-2016: Que Establece los Requisitos y Especificaciones de Sustentabilidad de Calidad de Playas. Norma Mexicana, 2016. www.gob.mx/cms/uploads/attachment/file/213866/NMX-AA-120-SCFI-2016.pdf. Accessed on: 19 Mar. 2019.
[24] Segrado, P.R. & Arroyo, L., El Método de la Capacidad de Carga Turística Aplicado a la Medición de la Sustentabilidad de Cozumel, México. *TURyDES*, **2**(5), pp. 1–10, 2009. Available from: http://www.eumed.net/rev/turydes/05/sa.htm. Accessed on: 23 Nov. 2017.
[25] Ley General del Equilibrio Ecológico y la Protección al Ambiente. Cámara de Diputados del H. Congreso de la Unión, 2018. www.diputados.gob.mx/LeyesBiblio/pdf/148_050618.pdf. Accessed on: 10 Apr. 2018.

NEED FOR INNOVATION IN TOURISM DEVELOPMENT OF THE LANDSCAPE DURING THE MATURE STAGE OF THE DESTINATION: PIE DE LA CUESTA, ACAPULCO, MEXICO

KARLA ROSALBA ANZALDÚA-SOULÉ[1], MIRELLA SALDAÑA ALMAZAN[2], MA. ELVIA CHAVARRÍA SOLÍS[1] & CESAR DANIEL AGUILAR BECERRA[1]
[1]Facultad de Turismo, Universidad Autonoma de Guerrero, México
[2]Centro de Ciencias de Desarrollo Regional, México

ABSTRACT

The district of Acapulco de Juarez in the state of Guerrero, Mexico has experienced a growth in tourism infrastructure that is reflected in its different stages of modernization (traditional or nautical, golden, and diamond). Urban evolution has brought impacts in the form of landscape alterations, social tension, and pollution amongst others. The Sectorial Plan for Tourism Development of the Metropolitan Area of Acapulco recently incorporated a fourth zone: Pie de la Cuesta that protects an environment of tranquility only some 10 km from the traditional area. Recent studies of the landscape and tourism enabled the revision of development, from a sustainable perspective, of elements from the landscape that hold an attraction motivating tourist displacement. The report mentions both considerations, valued in the diagnostic results based on Butler's Life Cycle theory, where the destination is reported as being at a mature stage with evidence of decline. The objective of the investigation derived from the article is the evaluation of landscape resources at Pie de la Cuesta with the aim of maximizing tourism potential to strengthen the services offered through new activities, without reducing resource quality in the long term.
Keywords: landscape, tourism development, Life Cycle, Pie de la Cuesta.

1 INTRODUCTION

The district of Acapulco de Juarez represents the 2.95% of the area with the highest growth of touristic activity in the state of Guerrero [1]. Gómez and Aldama [2] remind us that in the 1950s it was a small, quiet village full of traditions, with a strong cultural identity and an impressive natural environment, while for the 1970s, Acapulco was known nationally and internally as a tourist destination.

This expansion increased until the 1980s, which brought with it continued urban growth which was the object of tidying up at various times, giving as a result an uneven spread of services in distinct areas, favoring those needs of mass tourism. Among the effects of said transformation are alterations to the countryside/landscape, social tension, and pollution, a consequence of unplanned growth [3].

As Gordon [4] mentions, this type of tourism has a principal characteristic – unequal growth, as reflected in the infrastructure concentrated around the urban center: Acapulco. The city boasts 42% of hotels, 63% of hotel rooms, and the largest earnings through tourism in the whole state of Guerrero [5]. Tourist activity is the most important to the districts economy and is concentrated primarily in and around Acapulco Bay, the beaches of Island de la Roqueta, la Quebrada, Puerto Marquez, Pie de la Cuesta and the historic fort of San Diego.

A natural consequence of the bifurcation of the peninsular and the Coyuca Lagoon, Pie de la cuesta maintains a peaceful atmosphere only 10 km northwest from the traditional zone and

WIT Transactions on Ecology and the Environment, Vol 248, © 2020 WIT Press
www.witpress.com, ISSN 1743-3541 (on-line)
doi:10.2495/ST200171

nautical area of Acapulco's port. Its main access is from the Federal Acapulco-Zihuatanejo and Mexico-Acapulco highways [6].

Pie de la Cuesta lacks, among others, the infrastructure for mass tourism. Affluence figures and hotel occupancy rates are not registered in divided terms as is the case in the rest of the tourist areas in the district [7]. Only recently was it incorporated as the fourth zone by the Sectorial Plan and Tourism Development to the metropolitan area of Acapulco.

While the Tourism Secretary's page recommends Pie de la Cuesta as a place to appreciate a sunset it lacks better information about the natural resources that include – the 6 km extension of beach territory, its strong waves favored by expert surfers and admirers of the imposing Pacific Ocean and the recreational activities on offer around Coyuca Lagoon that belong in part to the districts of Acapulco and the adjacent Coyuca de Benitez.

These resources highlight natural, cultural, and historic characteristics that are partially taken advantage of by tourism service providers forming an incipient compliment offered to the visitors of Acapulco. In addition, the beach, surf, sunsets, and lagoon offer a cultured landscape that distinguishes Pie de la Cuesta from the rest of the municipality.

The cultural landscape establishes a study subject that is represented as a functioning site of interaction between people and their environment, both maximizing the activity steadfastly associated with society: tourism. The local inhabitants are involved in this evolution of tourist activity, giving it meaning and uses as well as change, also brought about by the visitors.

2 SOCIAL AND TOURIST ANALYSIS OF PIE DE LA CUESTA

The municipal government of Acapulco de Juarez plans tourist activity by means of the area: Traditional, Golden, Diamond, and the fourth zone Pie de la Cuesta [8]. The latter is a recent inclusion, and as a result, actions are unclear or in a stage of proposal. An example of this is the failure to pursue the concerns voiced by the Propietarios de Negocios y Avecindados de la Franja Turistica, a workshop offered by the Promotion and Impulsion of Competition Commission [9] which seeks to obtain information on service providers with the aim of analyzing the situation and visualizing it for the future, for example proposing alternatives for a better development of tourist activity. Said workshop is part of the Programa Nacional para el Desarrollo y Fortalecimiento Empresarial (National Business Development and Fortification Programme).

The proposals and unrest of association members, in turn, incorporated new activities that generated special interest among visitors by enjoying places built during the long history of the community: requiring the unification as mentioned by Álvarez [10] constructing an agreement between the interests of the while trying to conserve authenticity, as a system of resources and tourist attractions that transcend the real worth as one attraction more. Also, in the case of Pie de la Cuesta, a guaranteed preservation of the countryside and its identity is required as part of essential tourism development. Such a guarantee provides elements to differentiate the service and the supposed projects of the communities that possess a strong sense of identity and their own cultural worth, legacy, and lifestyle. They are then better placed to fulfill the development of tourism formed by their own sense of worth in the community and integrating attractions for potential tourists [11].

The previous is relevant with regard to social, economic, and environmental problems surrounding Acapulco and the results diagnosed based in Butler's Life Cycle theory where the destination is reported in an advanced stage with signs of decline [12]. In this way, valuing the country-side as a heritage for all, according to the Corporacion Patrimonio y Paisaje de Chile (The Countryside and Heritage Corporation of Chile) [13], facilitating the sustainability of social initiatives to improve the quality of life, increasing the possibilities for better behavior and stimulating a sense of belonging, dignity, and pride to live there.

The lack of clarity within community efforts increases the shortfall of data registry of visitor satisfaction and hotel occupancy in the tourist regions. Such a shortage of information could lead to different impacts in at least two different ways.

1. The promotion of Acapulco as a destination, where Pie de la Cuesta is aforementioned as a place to enjoy sunsets, where the official page of Acapulco's Municipality neglects to mention the existence of 35 hotels and 45 restaurants that make up the place: ignoring distinctive parts that could attract new demand. An example can be seen in Fig. 1 – the Coyuca de Benitez Lagoon, a swamp/mangrove ecosystem that measures 10 km in length east to west and more than 5 km wide, flowing west in a channel to the mouth of the river with the same name. Coyuca Lagoon is a central element to the population; well known by professional water-sports enthusiasts for skiing, kayaking, and rowing amongst others [6]. It's also home to endemic species incorporated into legends and cuisine by locals that shelter and the islands of Montosa and Presido. Here boat tours are given while guides tell stories of ancient families that used to inhabit the islands. (One of which was once used as a provincial prison.)
2. In Table 1, which relates to the correlation of available rooms and their occupation from 2015 to 2018 [14], the services maintained growth despite occupancy not recovering. From the other three destinations that surpass Acapulco in the offer of available rooms, only Mexico City shows a marginal positive variation, while Cancun and the Mayan Riviera present distinctive stages and implemented strategies to achieve development of their areas.

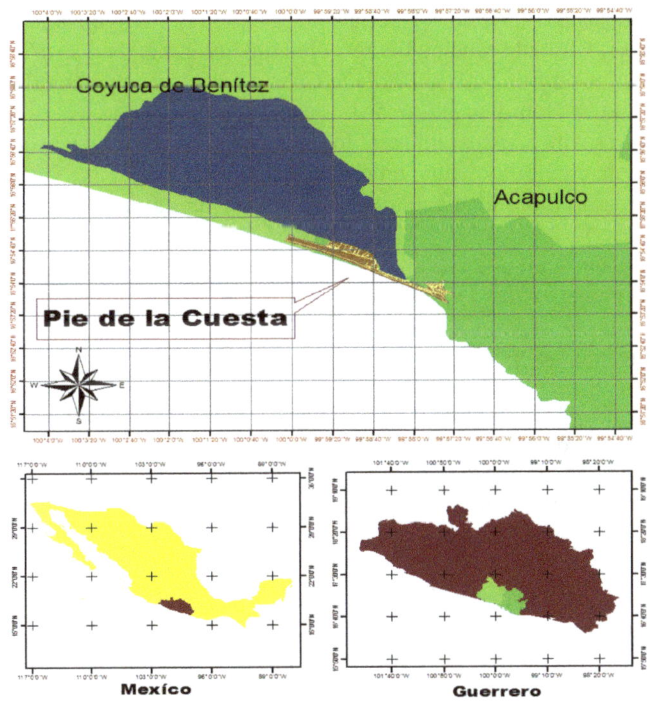

Figure 1: Study area.

Table 1: Hotel activity in selected tourist centers. *(Source: data from the hotel monitoring system DataTur from the Federal Government's Tourism Secretary.)*

Hotel activity in selected tourist centers												
Tourist destination	Average available rooms						Occupied rooms					
	2015	2016	2017	2018	Variation 15–16	Variation 17–18	2015	2016	2017	2018	Variation 15–16	Variation 17–18
Riviera Maya	36,524	37,336	41,554	43,598	2.2%	4.9%	29,699	29,950	34,234	34,740	0.8%	1.5%
Cancún	31,815	31,662	33,486	34,922	–0.5%	4.3%	24,520	24,315	25,707	26,065	–0.8%	1.4%
Acapulco	18,591	18,573	18,827	18,972	–0.1%	0.8%	7,444	7,473	8,188	8,847	0.4%	8.1%
Ciudad de México	49,819	51,777	51,420	51,344	3.9%	–0.1%	33,438	34,548	33,756	33,453	3.3%	–0.9%

On the other hand, there's no record of the number of visitors who use the condominium and second residences offer, which in the case of Acapulco is also vital information, because it is very handy for national tourism's accommodation. Both the increased offer of available rooms and second residence favor the motel of micro tourist sites, places where tourists find all the services, without contact with the local population that is segregated.

With regard to these previous considerations, the principal question that the investigation seeks to answer is if the natural, cultural and historic characteristics that possess the resources making up the landscape of Pie de la Cuesta could be used to strengthen the existing services with new activities, without reducing the resource quality in the long term?

For this reason, we try to explore Pie de la Cuesta's landscape potential as part of its real worth, as described by Bourdieu [15, p. 54].

"Certain features which appear inherent to the person as well the agent, such as authority, prestige, reputation, credibility, fame, notoriety, honor, good taste, etc. Understood like this, symbolic wealth is no more than economic or cultural capital in how much is known and recognized".

In this manner, the symbolic wealth contributes to the development of the area more than proportioning an image to brand the region, towns, areas and countries, surpassing Álvarez's comment in relation to tourism's contribution to the image of these distinct areas twofold; the general image and the specific image from tourism development [10].

3 THEORETICAL CONSIDERATIONS FOR DECLINE OR REJUVENATION

Theoretical planning, open to continuation is useful in order to outline topics such as the development of the landscape as symbolic wealth; sustainability (environmental, economic, and social) and in addition contributing both empiric and theoretical elements that permit investigation and intervention within a mature destination.

In the investigative part of the hypothesis, one of the instruments used to bring about sustainable development is the analysis or model – The Cycle of Life, proposed by Butler regarding tourism. This theory seeks to analyze the behavior of demand, the carrying capacity of the tourism centers and tries to forecast the positive relationship between the increase in visitor numbers and tourism development. Butler [16] predicts an evolution in the development of a destination where it is possible to follow up to six stages or phases: discovery or exploration, beginning or implication, development, consolidation, stagnation, and post stagnation.

The original Life Cycle model was inspired by ecological models, also mentioning the word "evolution" with respect to the problems of tourism [17] for which the use of said term has been criticized for the comparison of centers to living entities and their capacity for change. However, the term is used to refer to the process of change, apparently natural or out of control, as foreseen in the original model, thus showing a way to survive in a competitive environment [18]. Butler [16] maintains as a principle argument in his theory that if a tourist center is not constantly changing, it is open to financial obliteration (economic death) and in his article "Implications for Management of Resources", reveals that those tourist centers not taking measures to adequately manage their resources to maintain competitive in the market, don't progress and will only have the deliberate actions of people, which is to say that those who don't achieve adequate intervention can provoke a stage of decline in the destination.

Criticism of the first theory created by Hovinen [19], attributed to Wall [19] focused on the conception of the carrying capacity of the tourist areas and the application of the development cycle. Other investigators have championed the for the rejection of the Life Cycle for Tourist Areas (CVAT) such as Leiper [19] and Aguiló et al. [20] arguing that such a model is purely theoretical.

Confronting these opinions, Getz [21] supports the theory and explains that the relevance resides in tourism planning, while Prosser [19] arrived at the conclusion that the criticism formed against the model is due to its complicated concept, despite it still being used in its original form and offering perspectives in development. Papatheodorou's case [22] holds that the development and competition within a tourist destination changes greatly between individual companies at distinct levels, and that this relates to changes in preferences and tastes of the visitors. For this reason, the policies of managing demand in the Life Cycle should be planned in an integrated manner to promote the service(s) on offer. Following the same route Coles [23] placed the Life Cycle analyses in a sales context related to marketing models. This showed that criticism, in particular, referred to the lack of stage definition, a lack of validity for all classes and forms, that the products are live beings and that the participation of this in any one moment could be suitable. What's more, that the CVAT implies that the nearing of the final phase of any destination is accredited to a perceived reduction in attractiveness on the part of the customers or its inability to remain competitive on a local, regional, or global scale. For this reason, it's additionally important to pay close attention to the opinions of local residents in particular in a political context, as highlighted by Gale and Botterill [24].

During a destination's life cycle, apart from changes that occur in the number of visitors or in the composition of tourists and travelers in relation to the costs and benefits offered, the vulnerability of the heritage is also changed which allows the beginning of sustainability inside the destinations. Actually this theme is taken into account in the development of tourist destinations at local, municipal, state, national, and worldwide levels, owing to the high state of degradation suffered by resources as a result of not being developed in a balanced manner.

This inadequate use of the environment causes the site to advance rapidly through what is referred to as the distinct phases of the life cycle until a stage of decline is reached. Such is the case of Acapulco which is/was analyzed using Butler's model for the years 1945 to 2005, in which geographical, demographical, urban, and touristic changes were studied along with the development of the different zones (traditional, golden and diamond). The analysis results found that the port is currently in a phase of maturity with respect to the Destination Life Cycle, with indications of decline [12].

Priestley and Mundet [25] proposed adding a restructuring stage after the stagnation phase of the site, with the aim to prolong its life, while Baum [26] proposes including a reinvention stage during the final declining phase, promoting a change in the product and an economic, social and environmental balance at local, municipal and state level. In the same way, the National Development Plan [27] promotes local tourism development in the areas and municipalities of Mexico with the aim to harness natural resources with strict care being taken to conserve the environment, while designing strategies that generate high-level social benefits in areas of tourism potential.

In this manner of thinking, the landscape is the major resource of most importance to tourism now that it's of particular interest for the activity and can also be taken as part of three dimensions: cultural, eco-geographical and visual aestheticism [27] so much so that one of the first things to determine which elements of the landscape were to be incorporated corresponds to definitions created by Benseny [28] such as elements that possess high value and motivation for tourist displacement, including in equal measure cultural and historic resources that make up the area. Generating tourism development of said elements in Pie de la Cuesta with the objective of helping to counteract the decline of Acapulco, requires the structuring of activities that manage to transmit aesthetic, geographical, historical, and cultural characteristics from a sustainable perspective. Successful cases based on this approach include; The Belgian Landscapes (Saint Nazaire and Keroman Base in France; Atlantik Wall, in Denmark; Karlkriese and LichterfeldeSud, in Germany) and also Gaucho's Way (that extends from Laguna in Brazil to the Silver Sea in Argentina) amongst others. Both cases achieved the revaluation of their resources through a cohesion in a community aware of its value and responsibility in protecting it, in order to promote the areas of potential tourism to local and regional levels [29].

Investigations into destination, tourist area, and product complexities and/or problems that apply the CVAT concept, all relate to a fundamental element in common – "control", however all differ in the levels of its handling. Clear examples are the tourism destinations which owing to them being made up of many elements of different sizes, ambitions, and approaches, reflect a notable lack of capability to coordinate, while in the case of the product through changes and improvements that are easy to obtain.

In the case of a destination in decline, according to Brougham and Butler [19], the interest in new development is not found in the tourism center but in its surroundings that share characteristics similar to the original state of the area. For this reason, numerous applications exist for the tourism area's Cycle of Life from different perspectives, despite tourism being an activity that by its very nature is in a state of constant change. Kapczynski and Szromek [30] state that Polish water resorts in the last half-century have followed the life cycle stages and managed to overcome the decline phase and enter an expansion stage in the last few years. Cohen-Hattab and Shoval [31] indicate that the cause of a destination arriving at its final stage could be the result of failure to follow policies established during planning, which is demonstrated in their investigation of Israel. In the context of tourism and wine, Tomjenovic and Razovic [32] examined the Life Cycle concept in the development of Croatia's tourist regions, with the innovation of incorporating the ideas and attitudes of the owners of wine warehouses. Whitfield [33] applied the model in the Tourism Convention in the UK, focusing on the use of renewable activities for rejuvenation and concluded it could be applied to tourism centers. In the case of Duffus and Dearden [19], the model was applied to illustrate how changes in demand could affect specific products and help the management of wildlife resources.

The case of Las Vegas represents a destination depending on the marked stages of the life cycle. Today should show a decline, however its constant renewal and renovation could place it as a representative model to prolonging life. However, Las Vegas can't represent a central model as its development is based on cunning sustainability.

After the numerous applications mentioned above, it is impossible to leave to one side the debates that have sprung up surrounding the accuracy of the Life Cycle in the development stages of a mature destination. Agarwal [19] proposed an additional phase of restructuration or reorientation which is added to the stagnation period. From the same perspective, Priestly and Mundet [25] also discuss the need to add it after the stagnation phase to indicate that they are targeting an ever-more sustainable market, only that the lengthening of the life cycle depends on development time and the existence of a market for local activities. If this period of reorientation wasn't accepted, then it would have to be placed at the final stage and named reinvention, consisting of a complete change to the service [26]. However, although this wouldn't avoid, after some time, the arrival of the period of decline, now the issue is that if the destination is regenerated quickly, it will have an existence that much shorter [34]. To avoid this, it is suggested that the application of indicators helps to predict stage movement which in turn helps to take measures to prolong the life of a destination. In order to understand the development of a destination, according to Hall [19] it is necessary to understand the life cycle and tourism in general. A theme that has been of interest in tourist areas in recent years is the changing of attitudes of local residents [19], as they also participate in the development of tourism, causing changes without giving them consideration [35]. Other players who also participate in destination development are the tourists themselves, who facilitate ways to obtain what's needed, now that tourism has become fragmented, tourist centers have become more complex and predicting phases of their development more difficult. For this reason the prediction application has been shown by Berry [36] and on the same line Manente and Pechlaner [37] have integrated an early warning system, however, if the predictions are not accurate it won't serve for much. Liu et al. [38] observed that actually few destinations have carried out profound research and fought to prolong their life cycle like Hawaii [38], which is an area of multiple destinations that can avoid arriving at the rejuvenation stage, thus avoiding decline. However, even when existing investigations are complete and surrounded by policy, their application is frequently ineffective [19].

However, there are studies that recognize that even though each beach tourist destination stands out for its economic activity based on the exploitation of natural, cultural and historical resources, there is a certain relationship between competitiveness and quality of life. Some examples are: the beach destinations Puerto Vallarta and Mazatlán who presented some lags, while Acapulco ranks as the most laggard and least competitive. It seems that the environmental and social exploitation of traditional tourist destinations such as Acapulco, shows an exhaustion of the dominant tourist model based on the use of natural, cultural and historical wealth. What seems that as an alternative to prolong the life cycle of a destination it should be developed with greater inclusion, well-being of the communities and that is considered sustainable [39].

4 CURRENT AND FUTURE CHALLENGES OF TOURISM

The Acapulco tourist destination, recognized worldwide, grew in an unplanned way, in the heat of the fashion for sun and beach tourism focused on international tourism, imposed around the 1940s [40]. Currently, it has become a national destination with an offer for all socioeconomic levels.

The evolution of the Acapulco life cycle has not been positive, since it is in the stage of maturity with a tendency towards decline [41]. What is perceived at this stage is their loss of competitiveness for various reasons such as serious environmental and social problems. In the short term, there are no strategies' insight to revive this most important sun and beach destination in the country and in Latin America [42].

In this sense, Pie de la Cuesta can contribute to strengthening the tourist offer of Acapulco, promoting a balanced use of the landscape, generating tourist activities compatible with long-term development, in order not to reduce the quality of the landscape resource and promoting a new stage of rejuvenation for Pie de la Cuesta and against the decline that Acapulco is experiencing.

Some key aspects to promote the healthy tourist development of Acapulco are the use of the main emblematic landscapes of the Pie de la Cuesta area (the sunset and the Coyuca de Benitez lagoon) that must be done from a low perspective. Impact, since the area, does not have a regulation of uses and restrictions for tourists and even for service providers themselves, so it becomes essential to correct or take measures as soon as possible so that they are not demerited.

The main tourist receiving the Pie de la Cuesta area is national, local, and incipiently international. National tourism is the one that makes the highest tourist spending, followed by international and finally local [43]. It is evident that the national market must be analyzed to know the main tourist demands and seek to make the use of resources compatible in a balanced way.

The challenge will be the creation of tourist activities related to the three dimensions of the landscape (cultural, ecological-geographic and visual aesthetic) for; Strengthen the current tourist offer of Pie de la Cuesta, generate jobs and reinforce the self-esteem of the community, helping to value their beliefs and customs so that they shape a better tourist future and do not make the same mistakes as in tourist areas (Traditional, Gold y Diamond) of the port of Acapulco.

5 CONCLUSIONS

In synthesis, the suggestions and concerns of the Asociation de Propietarios de Negocios y Avencindades de la Franja Tourista (The Association of Business Owners and Neighbourhoods of the Tourism Strip) members need the integration of customer satisfaction registry and an appraisal of the resources that make up the cultural landscape. Such an integration is needed to form new scenes of development merged with the interests of the community. Patronizing new offers of services and renovating existing activities is the rejuvenation that would allow Pie de la Cuesta to survive amidst the declining environment of Acapulco. On the other hand, should the current political will to incorporate it as a newly installed tourism area continue to cause increases in hotel numbers (and consequently occupancy) thus losing beach and coastal views, it will be orchestrating the demise of one of the few original attractions that remain in the area. Increased awareness of the value of their natural landscape from local residents during this phase is one of the strengths that can redirect or reposition the cycle. Overcoming the decline of Acapulco as a tourism destination is urgent for the economy of Guerrero with measures emerging immediately for the inhabitants around the municipality.

Acapulco as a tourist destination in a phase of decline must take advantage of strategies for the use of areas with sustainability criteria. In which actions that achieve significant improvements in the conservation of natural resources and valuation of culture and history are constantly carried out to strengthen the current offer and plan future tourist activities.

It is urgent to start combating social, economic, and environmental problems such as public insecurity, high levels of marginalization, educational backwardness, and environmental contamination that for years has continued to increase and consequently shores the tourist destination to the stage of decline.

REFERENCES

[1] Guerrero State Government, *Encyclopedia of the Municipalities of Mexico*, National Institute for Federalism and Municipal Development, 2015.

[2] Gómez, S. & Aldama, M., *Modernity and Tourism: Repercussions in the Case of Acapulco*, Pandora: Guadalajara, Jalisco, pp. 35–39, 2007.

[3] Fascicle 3, Secretary of Tourism. www.sectur.gob.mx/work/models/sectur/Resource/14774/fasciculo3.pdf. Accessed on: 24 Sep. 2018.

[4] Gordon, B.M., *Contemporary History: Mass Tourism: A Problematic Concept in the History of the 20th Century*, pp. 125–156, 2002. https://addi.ehu.es/handle/10810/37937. Accessed on: 24 Sep. 2018.

[5] Statistical Compendium of Hotel Activity, Ministry of Tourism. www.datatur.sectur.gob.mx/ITxEF/ITxEF_GRO.aspx. Accessed on: 17 Oct. 2018.

[6] Guerrero State Government, Beaches in Acapulco, Acapulco Guerrero. Accessed on: 18 Sep. 2018.

[7] Cienfuegos, M.N., Personal communication, 18 Sep. 2018, Business and Neighborhood Owners of the Tourist Strip of Pie de la Cuesta, Acapulco, Guerrero.

[8] Tourist Documents Center, Tourism Development Sector Plan for the Metropolitan Area of Acapulco. Disponible en [1CD-ROM], México DF. Accessed on: 15 Nov. 2018.

[9] Commission for the Promotion and Promotion of Competitiveness, National Program for Business Development and Strengthening, Pie de la Cuesta, Acapulco. Accessed on: 20 Nov. 2018.

[10] Álvarez, A., The contribution of tourism to the comprehensive development of receiving societies. *Theoretical-Methodological Aspects. Politics and Society*, **42**, pp. 57–58.

[11] Dean, C. & Northcote, J., Social capital, tourism and regional development: SPCC as a basis for innovation and sustainability. *School of Social Sciences and Humanities. Murdoch Research Repository*, **7**, pp. 502–522, 2004.

[12] Bergeret, M.R.J., Garay, V.A. & Quintero, L.A., *Acapulco Maduro de Litoral Tourist Destination: Critical Analysis of the Evolution of the Acapulco Tourist Destination from 1945 to 2005,* Autonomous University of Guerrero: Mexico, pp. 11–28, 2011.

[13] Heritage and Landscape Corporation of Chile, The purpose of this corporation. www.patrimonioypaisaje.cl. Accessed on: 20 Oct. 2012.

[14] Secretary of Tourism, Hotel Activity. www.datatur.sectur.gob.mx/SitePages/ActividadHotelera.aspx. Accessed on: 17 Mar. 2012.

[15] Bourdieu, P., *Pierre Bordieu y Capital Simbolico*, Ideas: Madrid, p. 54, 2003.

[16] Butler, R.W., The concept of the tourist area life-cycle of evolution: implications for management of resources. *Canadian Geographer*, **24**(1), pp. 5–12, 1980.

[17] Ravenscroft, N. & Hadjihambi, I., The implications of Lamarckian theory for the TALC model. *The Tourism Area Life Cycle. Volume 2: Conceptual and Theoretical Issues*, ed. R.W. Butler, pp. 150–163, 2006.

[18] Ritchie, J.R.B. & Crouch, G.I., *The Competitive Destination: A Sustainable Tourism Perspective*, CABI: Wallingford, 2003.
Cooper, C., *Tourism Area Life Cycle*, ed. R.W. Butler, Goodfellow Publishers: Woodeaton, Oxford, 2011. www.ibtcl.co.uk/web_documents/toursim_area_life_cycle_original.pdf. Accessed on: 2 Feb. 2012

[19] Aguiló, E., Alegre, J. & Sard, M., The persistence of the sun and sand tourism model. *Tourism Management*, **26**(2), pp. 219–231, 2005.

[20] Getz, D., Tourism planning and destination life-cycle. *Annals of Tourism Research*, **19**(4), pp. 757–770, 1992.
[21] Papatheodorou, A., Exploring the evolution of tourist resorts. *Annals of Tourism Research*, **31**(1), 2004.
[22] Coles, T., Enigma variations? The TALC, marketing models and the descendants of the product life cycle. *The Tourism Area Life Cycle. Conceptual and Theoretical Issues*, ed. R.W. Butler, vol. 2, pp. 49–66, 2006.
[23] Gale, T. & Botterill, D., A realist agenda for tourist studies, or why destinations areas really rise and fall in popularity. *Tourist Studies*, **5**(1), pp. 151–174, 2005. http://tou.sagepub.com/content/5/2/151.short. Accessed on: 10 Feb. 2019.
[24] Priestley, G. & Mundet, L., The post-stagnation phase of the resort life-cycle. *Annals of Tourism Research*, **25**(1), pp. 85–111, 1998.
[25] Baum, T.G., Tasking the exit route: Extending the tourism area life cycle model. *Current Issues in Tourism*, **1**(2) pp. 167–175. www.informaworld.com/smpp/content~db=all~content=a908067201~frm=abslink. Accessed on: 15 Jun. 2019.
[26] Presidency of the Republic-Mexico, National Development Plan 2007/2012. http://pnd.calderon.presidencia.gob.mx/pdf/PND_2007-2012.pdf. Accessed on: 23 Oct. 2019.
[27] Benseny, G., El espacio turístico litoral. *Aportes y Transferencias*, **10**, pp. 102–122, 2006. http://redalyc.uaemex.mx/pdf/276/27610208.pdf. Accessed on: 1 Oct. 2019.
[28] Muñoz, A., Landscape assessment: An environmental management tool. *Chilean Magazine of Natural History*, **77**, pp. 139–156, 2004.
[29] Kapczynski, A. & Szromek, A.R., Hypotheses concerning the development of Polish Spas in the years 1949–2006. *Tourism Management*, **29**, pp. 1035–1037, 2008.
[30] Cohen-Hattab, K. & Shoval, N., The decline of Israel's Mediterranean resorts: Life cycle change versus national tourism master planning. *Tourism Geographies*, **6**(1), pp. 59–78, 2004.
[31] Tomjenovic, R. & Razovic, M., Wine and sun and sea tourism: fruitful relationship or imposible dream. *Tourism: An International Interdisciplinary Journal*, **57**(4), pp. 449–461, 2009.
[32] Whitfield, J., The cyclical representation of the UK conference sector's life cycle: The use of refurbishments as rejuvenation triggers. *Analysis of Tourism*, **14**(5), pp. 559–572, 2009.
[33] Di Benedetto, A. & Bojanic, D., Tourist area life cycle extensions: A step logarithmic approach. *Annals of Tourism Research*, **20**(3), pp. 557– 570, 1993.
[34] Russell, R. & Faulkner, B., Movers and shakers: Chaos makers in tourism development. *Tourism Management*, **20**, pp. 411–423, 1999.
[35] Berry, T., The predictive potential of the TALC model. *The Tourism Area Life Cycle Volume 2 Conceptual and Theoretical Issues*, ed. R.W. Butler, Channelview Publications: Clevedon, pp. 254–280, 2006.
[36] Manente, M. & Pechlaner, H., How to define, identify and monitor the decline of tourist destinations: Towards an early warning system. *The Tourism Area Life Cycle, Conceptual and Theoretical Issues*, ed. R.W. Butler, Channelview Publications: Clevedon, pp. 235–253, 2006.
[37] Liu, J.L., Sheldon, P.J. & Var, T., Resident perception of the environmental impacts of tourism. *Annals of Tourism Research*, **14**(1), pp. 17–37, 1987.
[38] Sandoval, P., Competitiveness and quality of life in tourist destinations in Mexico. *InterSedes*, **41**(20), pp. 1–21, 2019. DOI: 10.15517 / isucr.v20i41.38818.

[39] Camisón, Z.C. & Monfort, M.V., Repositioning strategies for mature tourist destinations: The case of the Costa Blanca. *Tourist Studies*, **135**, pp. 5–28, 1988.

[40] Solís, M.M., Hernández, L. & Villagomez, J., The perceived value of the tourist destination in relation to the profile of the tourist, Guerrero-Mexico. *Administrative Investigations*, **46**(118), pp. 1–28.

[41] Cárdenas, E., Acapulco, Guerrero through the centuries. *Magazine of the Faculty of Architecture of the Autonomous University of Nuevo León*, **10**(13), pp. 83–95, 2016.

[42] Anzaldua, K., Alcerreca, L., Avilez, H., Madrid, F., Conde, N. & Alpizar, V., The tourist use of the landscape from a sustainable vision; a proposal for Pie de la Cuesta, Municipality of Acapulco, Guerrero. Master's thesis, National Polytechnic Institute, 1, p. 233, 2013.

PARTICIPATORY RURAL PLANNING AND LOCAL CAPABILITIES TOWARD OFFERING NATURE TOURISM

ALMA R. BOJORQUEZ-VARGAS[1], ROMANO G. SEGRADO PAVON[2] & MARÍA DENY D. POOT QUINTAL[2]
[1]Universidad Autónoma de San Luis Potosi, Mexico
[2]Universidad de Quintana Roo, Mexico

ABSTRACT
In this study, ecotourism alternatives were identified that could complement the tourism offer in ejido Los Sabinos Numero Dos; which is a rural community located in the area of influence of the Biosphere Reserve-Sierra del Abra Tanchipa; this reserve is located in the Huasteca region of the state of San Luis Potosí, Mexico. The methodology carried out consisted of a participatory rural planning. Prior to that, a training workshop course was given for members of this ejido on conceptual issues and origins on nature tourism, sustainable development and the importance of land use planning. The participatory rural technique applied was participatory mapping for the identification of tourism resources with exploitation potential. The ejido members shared their learning and opportunities to open new activities as a new offer of nature tourism. As a result, new initiatives were identified in the form of local sustainable companies based on the potential to expand the tourism offer and allow greater benefits and interests of ejidatarios. It was found that the training course offered criticism, debates and reflections on the potential of its natural and cultural resources for tourism use. Likewise, participatory mapping led ejidatarios to share knowledge, analyse the location of interesting resources for tourism, their current conditions and the identification of new projects to expand the tourism offer.
Keywords: sustainable tourism, Sierra del Abra Tanchipa, community development, participatory Cartography, Huasteca Potosina.

1 INTRODUCTION

Among some data, the entry of foreign currency in Mexico by international travellers amounted to 1,941 million dollars, at the beginning of the year 2018 which represents an increase of 0.7% compared to the beginning of the year 2017 [1]. Regarding the number of tourists who entered the country at the beginning of 2018, it was 3.4 million; higher figure in 379 thousand tourists compared to the beginning of 2017, which means an annual increase of 12.7% [1]. The World Tourism Organization – UNWTO [2] indicated at the end of 2018 to Mexico as the sixth place in arrival of tourists, with the figure of 39.3 million foreign tourists, after France, Spain, United States, China and Italy in ascending order [2].

1.1 Increasing tourism in Huasteca Potosina region

This increase is encouraging but it also means a challenge of preparation, because it is not about increasing the tourist influx without taking into account adequate planning in infrastructure, training, service quality, much less, without the anticipation of social impacts and environmental that charge high bill to the natural and cultural heritage of the destinations.

The challenge is also reflected at the state level San Luis Potosí (SLP), and its constant growth of the estimated tourist influx of 7% per year in Huasteca Potosina region [3]. The arrival of visitors is constant most of the year, sharpening during Holy Week, summer vacations, festivities of the Dead (locally known as *Xantolo* Celebration) and, Christmas and New Year holidays.

The problem is suffered by both, local residents and tourists, who fail to fully satisfy their experience, given the great congestion of the most popular destinations. The director of Destination Management of the SECTUR of the last federal government period, said

doi:10.2495/ST200181

“problems of overloading by volume of tourists have begun to occur, especially in high seasons [...], which jeopardizes its viability as a tourist destination”, and He mentioned Xilitla, a municipality of San Luis Potosi State, as one of the five most visible and worrisome in the country [4].

What about countless natural and cultural resources with scenic beauty that could alternate the limited tourist offer currently, and in turn, distribute the mass of visitors? Entering the tourist activity is of interest to various locations, especially those with limited options in sources of economic income. The Huasteca region in San Luis Potosi has a wide variety of options that, if studied its tourist potential, could alternate the current tourist offer, since it has a diversity of sites with valuable natural, ethnic and archaeological heritage.

1.2 Ecotourism in the area of influence of the RBSAT

This document presents such concern, which allowed us to investigate the nature tourism alternatives around the Sierra del Abra Tanchipa Biosphere Reserve (RBSAT), located in the Municipalities of Tamuin and Ciudad Valles; the latter, as the main place of tourism companies in the region, such as hotels, restaurants and adventure tour operators.

Little more than ten years ago that the Los Sabinos Numero Dos ejido (Los Sabinos), located in the Influence Zone of the RBSAT (Fig. 1), has been driven in the activity of adventure and nature tourism, specifically, the use of an attractive natural underground cavity which impresses visitors by the formations of the speleothems inside. This cave, known as Las Grutas de Los Sabinos (Sabinos Grottoes) located in the ejido with the same name, had a certain boost with financial and training support from municipal governments between 2005 and 2009 [5] and despite this, and being located just 13 km from the city, it does not achieve significant benefits to detonate as a recipient of a significant number of visitors a year.

The RBSAT was declared a protected natural area on June 6, 1994 by the federal government as a presidential decree. In this reserve, a large area of sub deciduous jungle is protected, which are still in good condition. In addition, it is the little altered habitat of an important diversity of threatened flora and fauna and some declared endangered [6]. These extensions of protected forests make their ecosystems with biological diversity acquire recognition as a valuable natural heritage, not only regional, but national. Among the most prominent species are the jaguar (*Panthera onca*), the tiger or locally called tigrillo (*Leopardus Wiedii*) the green macaw (*Ara militaris),* the tamaulipeco parrot (*Amazona viridigenalis*), the chamal (*Dioon edule*), in danger of extinction and the palm soyate (*Beaucarnea inermis*) as a threatened species [7].

The polygon of the natural area protected (ANP) measures 21,464.44 ha and of that, almost 80% corresponds to the property of three ejidos (kind of collective land property in Mexico): ejido Las Palmas, Los Sabinos Numero Dos and el ejido Laguna del Mante; this last one mentioned occupies 15,737.68 ha [6]. On the other hand, only 425 correspond to ejido Los Sabinos Numero Dos; and despite this small space in the reserve, it is the ejido that has ventured into the tourist use of the RBSAT’s area of influence, due to the grottoes.

The purpose of this study was to apply participatory mapping tools to identify touristic alternatives in the ejido Los Sabinos Numero Dos, as a complement to the current tourist offer, since it only has tourism in the Grottos. Also, that alternatives should be low environmental impact, since it is located in the area of influence of the nature reserve. Therefore, local training on sustainability and nature tourism were both key elements in this study.

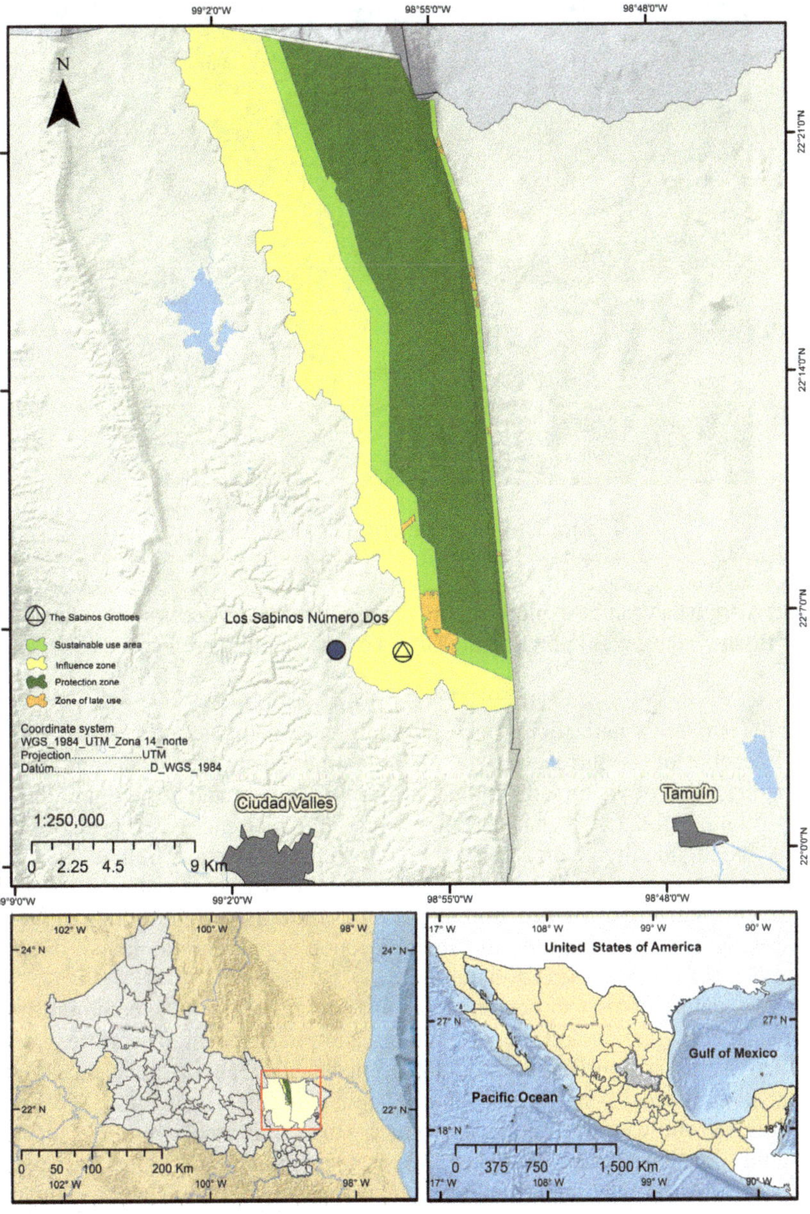

Figure 1: Location of the Los Sabinos RBSAT. *(Source: Bojórquez-Vargas et al. 2013 [5].)*

1.3 The Sabinos Grottoes

The type of soil and the geological characteristics of the area of influence of the RBSAT, offers underground cracks and narrow basements difficult to penetrate for the spelunking activity, also in the ejido Laguna del Mante, however, in the ejido there is a vertical cave called

Figure 2: Inside the grottoes Los Sabinos.

Sotano del Arroyo that can be explorable, but definitely for better accessibility, the main tourist attraction is where the caves known as the Grotto of Los Sabinos are located (Fig. 2).

2 THEORETICAL FRAMEWORK

The World Tourism Organization (UNWTO) defines sustainable tourism as "Tourism that takes full account of its current and future economic, social and environmental impacts, addressing the needs of visitors, the industry, the environment and host communities" [8].

Chung-Shing, Hoi & Lawal, based on many literature references, define nature tourism as "…that is based on direct enjoyment and observation of some relatively undisturbed phenomenon of nature or a range of nature dependent experiences with low environmental impact, labour intensive activities, and social and economical contributions" [9]. For this reason, seeking a balance between the social, economic and environmental dimensions is where tourism activity becomes sustainable.

In Mexico, nature tourism or ecotourism is strongly related to local communities, natural environments, conservation and proper planning and management. Therefore, "to qualify a tourism activity as ecotourism, it should participate and benefit the local communities, protect the natural ecosystem, follow sound planning and management, as well as target a group of tourists whose motives and behaviours are distinct from conventional tourists" [10].

When the development processes for a rural community or group are exogenously designed, local involvement in that development process is unlikely. That is, without interest, knowledge, or without active participation of one rural population, a local involvement in that development process is unlikely because it is usually considered from experts outside the community. This has been known for decades, however, the inertia of the alienated processes, decisions from outside, has touched many rural communities, one is the ejido Los Sabinos Numero Dos.

2.1 Participation in development processes

This is stated by Arnistein, quoted by Guillen et al. [11] when he argues that "citizen participation, when it is not pure cosmetic, is citizen power". Therefore, increasing participation implies redistributing power. For Arnestein, the ladder of participation classified the degree of the same as eight steps of an imaginary ladder in which the first steps consisted of non-participation, the pieces of the part of the ladder indicate a symbolic participation, while that the last rungs single out the true citizen power [11].

To achieve a participation and involvement of the local population in the development processes, that is, that participation of citizen power mentioned by Arnestein, where the self-management of the population is achieved, the process is supported by accompaniments based on methodologies and research techniques with participatory approaches, the facilitation or accompaniment of technicians or researchers will also not be without the consent of the same population.

It is then that, through community or local participation and focused on increasing their capabilities, when the process of empowerment of a community or a group of citizens is genuinely managed, community-based actions emerge from the bottom up. Only from top to bottom will scientists, researchers and the government be there to support and facilitate access to sustainable livelihoods and means of protection. Both stakeholder groups must also share knowledge [12]. However, many scientists and government officials often underestimate if they do not rule out, the value of local knowledge and community activities [12].

In that sense, Chapman and Kirk [13] argue that community participation is the mechanism for active community involvement in partnership working, decision making and representation in community structures. As Aref asserts "Without participation, there is obviously no partnership, no development and no program" [14]. The author also quotes to Miranda saying that "community participation rises the sense of control on the people over subjects that touch their lives and also promotes self-confidence and self-awareness". In that sense, is important to enable community members to become actively involved and to take responsibility for their own development and step by step improving their decision-making power [14], [15].

2.2 Social participatory mapping.

To begin, Osorio Campillo and Sánchez [16] define the cartography as:

"Elaboration of a map with various routes, lines, paths and paths fed from perceptions, events, knowledge, places of passage, images and conversations, which indicate the uniqueness of a territory. These paths are only readable on the map, oriented towards an experimentation that acts on the real".

These authors also quote to Alberich-Nistal when he says that the social map works as a tool that "allows us to see the relationships in the social network, in a more complete, complex and close to reality mesh, including hierarchical relationships or at different levels and also, the relationships between different networks" [16].

In congruence with the authors, the tool in principle is only descriptive but very useful for conducting participatory diagnoses on social networks; propose changes, new sets of action, develop possible scenarios, put yourself in a position of future of how you can work to change current relationships [16].

John Pickles, quoted by Perkins [17] acknowledged that:

"It is quite well established in critical studies (if not in practice) that the cartographic anguish of universalist modernist cartography has been practically suspended. Instead, there is a much more nuanced and multiform understanding of the practice and use of cartography, and in which the production of geographical images is understood as a completely social product".

However, it continues to be discussed and is continually divided between critical studies and the hegemonic practice of scientific mapping worldwide [17].

A wide variety of forms of participatory mapping have been used. To mention a few, Cadag and Gaillard [12] mention as examples: terrain mapping, “stone mapping”, sketch mapping, 2D scale mapping, GPS mapping and Geographic Information System (GIS) mapping Web based and interoperable; These are mentioned from the simplest to the most demanding in terms of resources. However, the least demanding and most accessible types of mapping (land and stone mapping) are often rejected as a cartographic technique because of the lack of credibility.

On the contrary, very sophisticated GPS and web-based GIS mapping require training and huge resources; In spite of this, they do not guarantee total credibility, since they can also be manipulated by people outside the community, despite being credible to government officials and scientists. On average, drawing sketch maps is a form of participatory mapping that is most frequently used, for example, for disaster risk reduction. However, since sketch maps are not scaled or georeferenced, they are often discarded by government officials and scientists [12].

3 RESEARCH METHOD

For this research, which aims to identify alternatives to expand the current tourism offer in ejido Los Sabinos Numero Dos, in congruence with activities of low environmental impact, a participatory methodology was chosen, since they are local residents and local stakeholders in this activity who know better the tourist resources they have. However, participants were required to acquire greater knowledge about the trends of nature tourism and the segments and activities in which they are classified.

To meet this need, before conducting the participatory mapping, a 16-hour workshop course was given to a group of the local population in August 2018, consisting of ejidatarios (people who belong to the ejido), ejidatarios’ children and local residents; who responded to the invitation stating that they were interested in venturing into the tourist activity.

The workshop course was held at the premises of the ejido house of Los Sabinos. The topics to be addressed and the time of the workshop course and what it would consist of were presented for consideration. Based on an agreement, the duration of the workshop course and the subjects of the same course were established in Table 1 as follows:

Table 1: Content of the topics to be developed in the Workshop Course.

Workshop topics
Topic 1. What is tourism?
Topic 2. Sustainable development
Topic 3. Sustainable tourism
Topic 4. Nature tourism
Topic 4.1. Ecotourism
Topic 4.2. Adventure tourism
Topic 4.3. Rural tourism
Topic 5. Community tourism
Topic 6. Importance of territorial planning

The topics were addressed using slides, reflection time and peer discussion and with the support of a printed manual for each participant, in which the workshop course content was developed in detail and with drawings and images and examples. Once the group of participants became familiar with the concepts, terms on sustainable tourism and the classification of activities in which nature tourism consists, the mapping was made up of work teams, with the active participation of ejidatarios and the facilitation of researchers.

4 RESULTS

In order to apply the knowledge acquired from the workshop course previously taught, the participants worked in small teams to trace the ejido territory and identify the tourist resources they consider to be exploited as nature tourism. To do this, flipchart papers, colourful markers and the guide of the printed course manual were used.

4.1 Organization of equipment for mapping participation

Some teams focused on identifying resources or activities with tourism potential in the ecotourism modality, others in the adventure tourism modality and others in the nature tourism modality. The facilitation of the team of researchers was conducted by guiding questions (Table 2).

Table 2: Guide for facilitation in participatory mapping. (*Source: authors' own.*)

Guiding Questions
1. What natural resources do we have in our territory?
1.1. What is needed (training, infrastructure, equipment) for its use?
1.2. In what period (short, medium or long) can you start taking advantage of?
1.3. Which external actors can contribute to its development?
2. What aspects of our cultural heritage can we share to tourism?
2.1. What is needed (training, infrastructure, equipment) for its use?
2.2. In what period (short, medium or long) can you start taking advantage?
2.3. Which external actors can contribute to its development?
3. What tourist attractions have been used in our territory?
4. Which of the above is wanted and can be started in the short term and with less complication? (Hierarchy of resources with tourism potential)
5. What activities and projects can we schedule to manage and carry out?

4.2 The role of researchers

The research team prioritized acting as facilitators and companions for the diagnosis and detection of resources with potential for tourist attraction. In this diagnostic accompaniment, the resulting maps and diagrams consisted of simple schemes that allowed us to geographically understand the environment, natural resources with potential for tourism, as well as local traditions and aspects of the cultural heritage of the population.

The main interest was that, through touring the parts of the ejido territory, tourism resources could be identified one by one and for this, the support of the manual that was used as a guide in the course, was a material that was constantly consulted in the equipment during the mapping.

This contributed to the discussion and recognition of what exists within the territory and of the possible activities and sites with potential to be offered as part of nature tourism; and, in that way, enrich the offer of activities that extend the stay of visitors who are interested in the Grottoes. That is, generate alternatives to what is currently offered, which will allow a greater economic spill, in addition to being able to offer food and locally made products have an interested market, tourists.

4.3 Enrichment and exchange of knowledge during participation

During the participatory mapping activity, a greater dialogue, a space for knowledge sharing, was observed in the participants. At some moments in the mapping, the young people recognized they did not know part of the territory that the elders described with great security and a sense of appropriation.

On the contrary, the youth identified areas of difficult access, ravines and rocky areas where it was possible to offer rappelling and rock climbing; both are adventure tourism activities that tourists who prefer the challenge in nature increasingly demand. This interest was observed in young people and they expressed interest in being trained; while the aged, identified activities of lower risk and more contemplation, to offer tourism (Fig. 3).

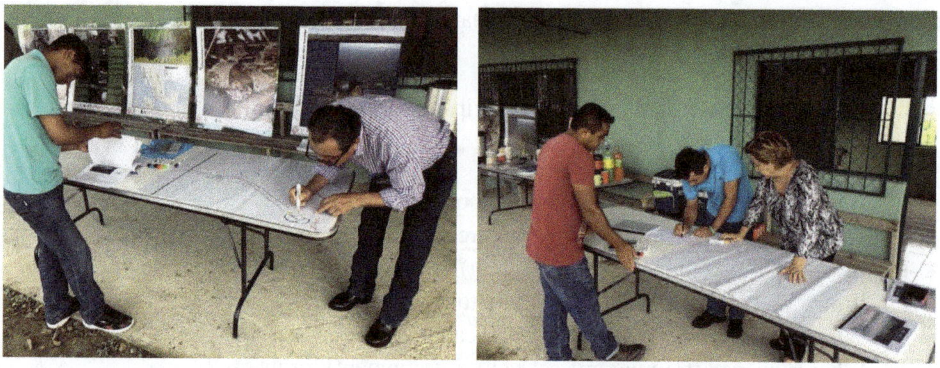

Figure 3: Identification of tourist attractions with participatory mapping.

4.4 Consensus agenda of projects and actions to manage

At the end of the presentation by each team, activities or sites with the possibility of being offered locally were listed. Those activities, sites and projects with greater feasibility of operating as part of the tourist offer that will complement the visit of the caves were prioritized.

Training and equipment needs, as well as identification of agents or support institutions for their development were listed and classified in the short, medium and long term.

The activities with the greatest interest in the short term were: (1) Establish an Environmental Management Unit (officially known as UMA) for the reproduction of native orchids for the purpose of environmental education, tourism and their trade, (2) Training in herbalism and elaborate products of the medicinal plants that can be used in the region, (3) Offer the traditional food and local cuisine to visitors, and, (4) To build a Bird watching tower by wood material and a zip line service over the Sotano del Arroyo (a natural abyss).

For this, training, equipment, support and financing needs were identified with respect to each service, product or project identified. It is then that, equally in a participatory manner, the governmental and non-governmental entities were identified to request the management of the same supports and the way they would be organized to carry out the management, assigning responsible in each identified need.

5 FINAL REMARKS

Thanks to the methodology used, it is considered effective and democratic for the diagnosis of alternatives in the tourist offer, which complements the current limited offer. It is reflected that offering only the visit to the caves does not encourage greater efforts to participate in other activities by the ejido. By identifying numerous activities and sites with tourism potential, as well as projects to develop tourism products and services, it boosted the motivation of the participants, who were surprised about the various options that can be offered locally for tourism, in addition to the caves.

The participatory processes developed in this research allowed to facilitate the interpretation, assimilation and understanding of the mapped information, making it visible and tangible for all participants. An opening was identified among the participants for sharing knowledge. It could be perceived that through the participation in rural diagnoses, together with the training of the economic activity of interest, in this case, nature tourism, it is possible to motivate and create local awareness about the potential of their territories.

Thank to this experience, it can be shared with the academic community and field extensionists for rural development processes, which participatory mapping provides to the interested parties, powerful means for the efficient management of land use, ways to identify alternatives of economic activity, and which serves as an effective tool for community organization.

ACKNOWLEDGEMENTS

Thanks to the willingness and active participation of members from Ejido Los Sabinos Numero Dos, during the work days and great availability the subsequent visits by researchers. Thanks to the staff of the Comisión Nacional de Áreas Naturales Protegidas CONANP, who officially lead the management program of the RBSAT and for their support for this work.

REFERENCES

[1] SECTUR, DATATUR, Resultados de la actividad turística 2018. www.datatur.sectur.gob.mx/RAT/RAT-2018-01(ES).pdf.

[2] UNWTO World Tourism Barometer – Barómetro OMT del Turismo Mundial 2018. https://doi.org/10.18111/wtobarometeresp.2018.16.5.1.

[3] Secture, Programa Sectorial de Turismo, Plan estatal de desarrollo del gobierno del Estado de San Luis Potosí 2015–2021 2015. www.slp.gob.mx/programassectoriales2016-2021/assets/E1_4.pdf.

[4] Vázquez, J., Revisarán capacidad de carga de Pueblos Mágicos. *El Economista*, 11 Jul. 2018. www.eleconomista.com.mx/estados/Revisaran-capacidad-de-carga-de-Pueblos-Magicos-20180611-0015.html.

[5] Bojórquez-Vargas, A.R., Hernández González, G. & Suarez Rodríguez, I., Aprovechamiento Turístico en Cuevas. El caso del Ejido Los Sabinos, Huasteca Potosina. *Turismo Espiritual II: Una visión Iberoamericana*, ed. R. Martínez-Cárdenas, Universidad de Guadalajara: Guadalajara, Jalisco, pp. 151–161, 2013. www.eumed.net/libros-gratis/2013/1238/indice.htm.

[6] SEMARNAT, Programa De Manejo Reserva de la Biosfera Sierra del Abra Tanchipa. Secretaría de Medio Ambiente y Recursos Naturales: México, pp. 1–200, 2014. www.sierradelabratanchipa.org/pdfs/Programa de Manejo RBSAT.pdf.

[7] SEMARNAT, Diario Oficial de la Federación, Norma Oficial Mexicana NOM-059-SEMARNAT-2010, Protección ambiental-Especies nativas de México de flora y fauna silvestres-Categorías de riesgo y especificaciones para su inclusión, exclusión o cambio-Lista de especies en riesgo 30 Dec. 2010. www.gob.mx/cms/uploads/attachment/file/134778/35.-_NORMA_OFICIAL_MEXICANA_NOM-059-SEMARNAT-2010.pdf.

[8] United Nations Environment Programme, Division of Technology and Economics, Making tourism more sustainable: A guide for policy makers 2005. www.e-unwto.org/doi/pdf/10.18111/9789284408214.

[9] Chung-Shing, C., Yan Chiu, H. & Marafa, L.M., The Mainland Chinese market for nature tourism in Hong Kong. *Tourism Geographies*, **19**(5), pp. 801–822, 2017. https://doi.org/10.1080/14616688.2017.1360387.

[10] Wondirad, A., Who benefits from the ecotourism sector in Southern Ethiopia? *International Journal of Tourism Sciences*, **17**(4), pp. 276–297, 2017. https://doi.org/10.1080/15980634.2017.1384133.

[11] Guillen, A., Sáenz, K., Badii, M.H. & Castillo, J., Origin, space and levels of participation. *International Journal of Good Conscience*, **4**(1), pp. 179–193, 2009. www.spentamexico.org/v4-n1/4(1)%20179-193.pdf.

[12] Cadag, J.R.D. & Gaillard, J.C., Integrating knowledge and actions in disaster risk reduction: The contribution of participatory mapping. *Area*, **44**(1), pp. 100–109, 2012. https://doi.org/10.1111/j.1475-4762.2011.01065.x.

[13] Chapman, M. & Kirk, K., Lessons for Community Capacity Building: A summary of the research evidence 2001. www.scothomes.gov.uk/pdfs/pubs/260.pdf.

[14] Aref, F., Sense of community and participation for tourism development. *Life Science Journal*, **8**(1), pp. 20–25, 2011. www.lifesciencesite.com/lsj/life0801/03_4106life0801_20_25.pdf.

[15] Herawati, A., Purwaningsih, A., Pudianti, A. & Surya, R.V., Rural tourism community empowerment based on local resources for improving community welfare: Case on Pentingsari village, Yogyakarta, Indonesia. *Review of Integrative Business and Economics Research*, **3**(2), p. 88, 2014. www.sibresearch.org/uploads/2/7/9/9/2799227/riber_b14-071_88-100.pdf.

[16] Osorio Campillo, H. & Sánchez, E.R., La cartografía como medio investigativo y pedagógico. *Dearq*, **9**, pp. 30–47, Dec. 2011. https://revistas.uniandes.edu.co/doi/pdf/10.18389/dearq9.2011.05.

[17] Perkins, C., Cartography: mapping theory. *Progress in Human Geography*, **27**(3), pp. 341–351, 2003. https://doi.org/10.1191/0309132503ph430pr.

CHALLENGES TO COMMUNITY PARTICIPATION IN HERITAGE TOURISM DEVELOPMENT: CASE STUDIES OF SHAHJAHANABAD AND NIZAMUDDIN BASTI IN NEW DELHI, INDIA

EKTA CHAUHAN
Brandenburg Technical University, Germany

ABSTRACT

In new Delhi, the capital of India, history and modernity exist side by side, making it an ideal tourist destination. The city is home to three UNESCO world heritage sites, 174 monuments of national importance and thousands of unlisted monuments. These sites exist not in isolation but are living heritage, where communities not only live around the structures but have made the structures part and parcel of their daily lives. Thus, tourism's impacts are most apparent at the level of the destination community and therefore it is essential to decentralize tourism development and policymaking. There is a need to integrate tourism development into the overall community-defined development goals, as is the case with several other developing nations, decision making is highly centralized in India and in Delhi, which results in low level of community participation, mostly at a coercive level and never at a consultative level. This research would, therefore, explore (a) local residents' perception of their role in tourism decision making and (b) barriers in the participation of local communities. The research was carried out through two case studies in Delhi: Nizamuddin Basti and Shahjahanabad. These areas are historically important and are the top tourist destinations for the city. The local condition of tourism development and community involvement, however, vary widely among them, making them ideal for the study. The study adopted a qualitative approach and used field observation and interviews to collect relevant data.

Keywords: community participation, cultural heritage tourism, sustainable tourism, participatory policy making.

1 INTRODUCTION

Cultural heritage has acquired a new dimension as a result of the recognition that the things it deals with, tangible or intangible, are beyond academic discourse. Cultural heritage provides emotional and intellectual support for communities which in turn helps them to establish their identity. If looked at as a driver of an individual or a community's destiny, heritage management must engage the people. There is a growing need, especially in developing countries, to recognise people's opinions and create more just processes in heritage management in order to provide for social sustainability as well as protect the relationship between people and heritage. Community participation thus becomes critical to the ethical performance of cultural heritage management [1].

While tourism is only one of the many functions of heritage, it has come to occupy a dominant role especially in economic terms as the world's largest and fastest growing industry. Tourism generates 10.4% of the global GDP and supports nearly 1 in 10 jobs [2]. As a result, tourism has also emerged as a crucial factor in planning and management of historic sites and landscapes. Potentially, tourism industry can play an important role in deciding the fate of the landscape as well the local community. Tourism, if managed sustainably, can make substantial contribution to poverty alleviation, conservation of the natural and cultural heritage, and to overall sustainable development [3]. This is especially the case in developing nations such as India, where natural resources and landscapes are

WIT Transactions on Ecology and the Environment, Vol 248, © 2020 WIT Press
www.witpress.com, ISSN 1743-3541 (on-line)
doi:10.2495/ST200191

abundant and relatively unexplored and where other activities have limited potential for sustainable socio-economic development.

As is the case with several other developing nations, decision making is highly centralised in India, and in the capital Delhi, which results in low level of community participation, mostly at a coercive level and never at a consultative level. The impact of tourism is felt directly and most intensely by the local communities. It has been argued that local communities must be involved in planning and management of tourism as they are best positioned to understand the tourism products associated with their social conditions and local culture. The knowledge of local tourism offerings and appropriateness to local situations signify host communities' involvement in tourism development [4]. In 1966, the World Bank classified the benefits of community participation in tourism into five types: enhancing insight, increasing commitment, improving ability, increasing resources, and social learning [5]. Hence there is a need to meaningfully engage local community in the decision making and planning of tourism process to maintain the right balance between the needs of all the stakeholders for long-term sustainable development of the historical sites and its surroundings.

Despite the agreement over importance of community participation in heritage tourism planning, there is a lacuna in knowledge especially in context in India. While there is an agreement that local stakeholders should be involved in decision making, there seems to be a limited understanding about the existing barriers that prevent the communities from taking part in the decision-making process. This paper would be looking at two heritage tourism hubs in New Delhi with the aim to (a) understand attitude of people towards cultural heritage tourism and (b) identify barriers to participation of local community in cultural heritage tourism management. This paper hopes to bridge this gap in academic research as well as assist policymakers in designing an appropriate framework for meaningfully enhancing participation of local communities.

2 LITERATURE REVIEW

2.1 Sustainable tourism

With the shift of development discourse towards sustainable development, studies have also started highlighting the concept of sustainable tourism with a focus on aspects such as harmful effects of mass tourism on resources of the local community (natural, built and socio-economic), role of tourism in promoting economic wellbeing, preserving natural and socio-cultural capital and achieving inter and intergenerational equity in benefits and costs of tourism [6]. Sustainability implies judicial use and resilience over time, so sustainable tourism would include optimum use of resources, including biological diversity; minimization of ecological, cultural and social impacts; and maximization of benefits for conservation and local communities. Sustainability also reflects the management structures that are needed to achieve the above. It is in the context of sustainable tourism that one has to view community participation.

People might travel for a wide range of attractions and pleasures and cultural heritage is one among many. Cultural heritage tourism or CHT has thus emerged an important segment of tourism industry that requires special attention due to its peculiar nature and resources. For the purpose of this study, the author would be using the following definition of CHT : CHT is the "movement of persons to cultural attractions in cities in countries other than their normal place of residence, with the intention to gather new information and experiences to satisfy their cultural needs and all movements of persons to specific cultural attractions, such as heritage sites, artistic and cultural manifestations, arts and drama to cities outside their normal country of residence" [7].

2.2 Community participation in CHT

Research has well established benefits of community participation in heritage and tourism management. There is, however, no overarching definition of "community" and the term is defined depending on the context. The World Heritage Convention promulgated by UNESCO in 1972 recognized cultural heritage as "a function in the life of the community" and the protection of World Heritage "is the duty of the international community as a whole to co-operate" [8]. The Council of Europe uses the term "community" to refer to people who "value specific aspects of cultural heritage which they wish, within the framework of public action, to sustain and transmit to future generation" [9]. It can be further said that people do not need to be in geographical proximity (of each other and the heritage) in order to form the community but should hold similar value towards the heritage resource and share a common goal and vision for its development [1].

There is a need for communities to be involved in local decision making especially for tourism as communities are deeply impacted by both positive and negative impacts of tourism [10]. Snyman argued that tourism should be community driven, where locals take the responsibility to control tourism infrastructures and facilities [11]. According to Haywood, "Healthy, thriving communities are the touchstone for a successful tourism industry". However, financial benefits and activities generated through tourism may not be compatible with a community's other social and economic objectives [12]. In fact the purely economic rationale of tourism development has been challenged and there is a demand for more responsive and community sensitive tourism [13].

There has been a discernible shift in the outlook for policy makers after the Earth Summit 1992. The declaration, Agenda 21, signed by 172 nations, endorsed "sustainable development", has since then become a guiding principle for planning process across the globe. Agenda 21, especially chapter 28, endorses a central role for local agencies in addressing sustainable development issues. Agenda 21 asks local authorities to adopt new ways of designing and implementing policies to adopt key ideas of sustainable development as well as incorporate a collaborative and democratic process of decision making actively involving the local communities [14]. Despite the fact that Agenda 21 makes only a scarce reference to tourism, its significance for tourism industry cannot be ignored.

2.3 Barriers in community participation for tourism

A number of barriers may exist that does not allow for realisation of full potential of community participation in tourism management:

2.3.1 Bureaucracy

A number of agencies from national to local levels are involved in tourism management. As tourism also directly impacts other sectors of the economy, there might be overlap and conflict of interests between different agencies as well.

2.3.2 Tourism planning is ad-hoc or absent

City level plans often do not even mention tourism as a separate activity and therefore do not put forward a strategic and integrated plan for its management. Tourism is seen as a purely commercial activity and collective action of private players (such as hotels, business, conventions) are considered enough.

2.3.3 Public participation may be seen as unnecessary, time consuming and utopian

There is no doubt that decentralised tourism management would accrue significant financial and executive costs. Executive class will have to dilute its own power, interact with citizens, educate them, and take on the danger of failure or bad publicity. These costs look significant at least in the short run

2.3.4 Conflict between profit motive and social goals

The tourism industry often may attempt to thwart any attempts that tries to incorporate social and environmental objectives as it is viewed as irrelevant or anti profit.

2.4 Knowledge gap

An understanding and knowledge gap exist between the government and community. Communities would require help in establishing a role for themselves which might be unavailable [12].

Tosun argues that while it is now established that community should be involved in development decision making and that currently there is only limited success of community participation, there has been no consensus as to what the reasons for this are. He, therefore, puts forwards three non-exclusive reasons for limitations to community participation in Tourism Development Process (TDP) in developing nations [15]. A summary of Tosun's three limitations are shown in Table 1.

3 METHODOLOGY

This research was carried out through qualitative interviews and field observation in the two case study areas. Social science researches have come to rely on case study approach increasingly and it has been recognised as an effective research strategy [16]. Using such an approach allowed to simultaneously consider multiple factors that influenced different community members such as individuals, family, organisations etc. in the selected case areas [17]. Given the complex nature of the research problem and multiplicity of actors and factors involved, case study approach was considered the best suited tool for this research.

A qualitative research methodology was chosen as it allowed the researcher to gather rich information from descriptive data sources such a government reports, archives etc. and from primary sources such as interviews and observation. By using less structured tools such as open-ended interviews and discussions, the participants were given the opportunity to come up with new ideas and build upon the researcher's knowledge rather than be restricted to pre-determined responses. The following two tools were used to gather primary data.

3.1 Participant observation

Participant observation has been used in a variety of social science disciplines as a tool for collecting qualitative data about people, process and cultures. For the purpose of this research, such an observation and participation enabled the researcher to study people and their activities and opinions in their natural settings.

3.2 Semi-structured interviews

Semi structured interviews were conducted using a-priori structuring of groups with theoretical sampling within the group to determine which case to integrate next. Thus, the structure of the groups to be interviewed was defined in advance by the researcher with a view of delimiting the totality of cases in such a way that all the cases can be integrated into the research [18]. Further to make each of the group representative, theoretical sampling as developed by Glaser and Strauss was used instead of statistical sampling. In theoretical sampling the process of data collection is controlled by the emerging theory. Thus, the data was simultaneously collected, coded and analysed to decide which data should be collected next and who should be interviewed. Sampling decisions were made in favour of individuals (or groups of individuals) who could promise to provide greatest insights in light of the material already used and knowledge drawn from it [19].

The research was carried out between December 2019 and February 2020. Several trips to research sites were made and meetings were held the with local community members which helped the researcher to realize the reasons behind community non-participation in tourism development. A list of questions was then prepared around the study theme for the interview script. In total, 47 in-depth interviews were completed.

Each interview was 35–45 minutes in length. All Interviews were recorded with the consent and later transcribed. Interviews were conducted in Hindi and English. Most of the local community members felt more comfortable and confident in Hindi and this ensured better expression and thus authenticity of the data. For internal consistency, interview responses were translated into the English language. Themes, sub-themes, and quotes were analysed to accomplish research objectives of the study. Reliability was attained using an interview script and matching data with field notes.

Table 1: Limitations to community participation in tourism management.

Operational limitations	Centralisation of public administration of tourism
	Lack of coordination among responsible authorities
	Lack of information
Structural limitations	Attitude of professionals
	Lack of expertise
	Elite domination
	Lack of appropriate legal system
	Lack of trained human resources
	Lack of financial resources
	Relatively high cost of public participation
Cultural limitations	Limited capacity of poor people
	Apathy/low level of awareness in the community

4 GEOGRAPHICAL SCOPE OF THE STUDY

This study adopts a case study approach and will be looking at two heritage tourism areas on New Delhi, India. While both the areas receive high number of domestic and international tourists every year, their conditions are vastly different which makes them ideal for a such a study. Though the findings of this study are limited to the case studies, they can be used to understand similar communities in other developing nations.

4.1 Nizamuddin Basti

Nizamuddin Heritage Precinct centres around the UNESCO World Heritage Site Humayun's Tomb along with the Hazrat Nizamuddin Basti and Sunder Nursery. The area is densely populated and is a tourist hotspot in the city. The Basti had 1500 households as of 2015 and a large floating population of tourists [20]. Along with Humayun's tomb which attracts domestic and international tourist, the area also houses the Dargah (shrine) of Saint Hazrat

Nizamuddin which is visited by pilgrims in high number. The area had however, fallen into decay in recent years. However, the situation improved once again with the launch of the Nizamuddin Urban Renewal Project in 2007 as a non-profit Public-Private Partnership between Aga Khan Trust for Culture and multiple government agencies that included the Archaeological Survey of India, Central Public Works Department and the Municipal Corporation of Delhi. The project was unique in its approach of combining heritage conservation with environmental and socio-economic development while working with local communities and stakeholders.

4.2 Shahjahanabad

Shahjahanabad or the Walled City of Old Delhi was founded by Mughal Emperor Shah Jahan in 1639. While the city survives even after 350 years, it is now reduced to a fraction of its original grandeur. Along with the UNESCO World Heritage Site Red Fort and Jama Masjid, the area is also home to several other monuments and structures along with a vibrant residential and commercial community. The area has transformed from a royal capital to a wholesale market with dense housing and depilated infrastructure. Despite this, it records a high number of tourist footfall every year; the Red Fort alone attracted 3.6 million visitors in 2018–2019 [21].

Shahjahanabad was also identified by Government of National Capital Region of Delhi to be nominated for inscription as UNESCO's list of World Heritage Cities (the nomination was later withdrawn due to political reasons but nevertheless highlight the importance of the area as a prime tourist location). Shahjahanabad future largely depends on its socio-economic revival and considering its history, the area's natural comparative advantage lies in sustainable cultural tourism. The essential precondition to the revitalisation effort requires a consensus among all stakeholders regarding the basic parameters under which Shahjahanabad is to be redeveloped [22].

5 FINDINGS

5.1 Institutional barriers

Given the current governance structure of heritage and tourism, local people had very few avenues to engage with the government and express their opinions. There are multiple agencies and jurisdictions involved in the management of heritage and tourism which leads to organisational obstacles. Even in cases where they were invited for public meetings and programs, there was no follow up. For example, even though a workshop for heritage homeowners in Shahjahanabad was organised by the North Delhi Municipal Corporation (NDMC) and Shahjahanabad Redevelopment Corporation (SRDC) [23], there was no follow up to ensure that residents' concerns raised during the meetings were addressed or that owners implemented the suggestions for the upkeep for their property. Such a lack of communication between the community and administrative bodies creates a feeling of mistrust among people which discourages them from participating in any future initiatives [24]. Another major barrier was the power disparities within the community. Marginalised and at-risk groups such as the poor, women, religious minorities and sexual minorities seldom got a chance to put forward their views. A section of the people also felt uncertain that their voices would be heard at all in the development process.

However, local residents of Nizamuddin Basti area were slightly more confident about their role in the development process. Aga Khan Foundation has been engaged in urban renewal and heritage management since 2007 in the locality. Community empowerment and consultation form an integral part of the project. Several local community-based groups now function in the Basti that allow people to engage with important issues such as health, education, sanitation and heritage management. To encourage local residents to contribute meaningfully towards tourism development self-help groups such as Insha-e-Noor and Sair-e-Nizamuddin were created. While the former engages women to produce and sell handicraft items, the latter trains youth as guides. This has successfully engaged locals as well as created a feeling of mutual trust among locals and administrative agencies.

5.2 Lack of awareness

It was observed that while most participants expressed a desire to be more intimately involved in decision making with the government, they did not know the ways to do so. One of the woman participant for Nizamuddin Basti said, "of course, we will go for whatever meetings the government call us for. It would ultimately affect us. I am sure I can manage household chores and the meeting". While such enthusiasm was welcoming, most of the participants did not know how to engage with the administration. There was a lack of knowledge regarding rights of citizens, governance, administrative laws and judicial system. They felt that local agencies should organise awareness and education drives.

There was also a lack of knowledge regarding the long-term impacts of tourism on local economy and society. While participants were eager to showcase their local culture to tourists, especially foreigners, they did not know the positive or negative impact that might accrue. A resident of Shahjahanabad said, "the level of awareness regarding heritage and tourism is very low. People know that tourists come here, appreciate the local heritage and buy local products. I don't think anyone understand how this affects their life".

5.3 Limited capacity

A local of Shahjahanabad said that the community still looks at tourism as a secondary activity for the area rather than the economic mainstay. He pointed out that the area is a now a bustling commercial centre with high rent and in such conditions, no one is willing to restore his property for tourists. It was observed that only limited opportunities were available for the locals to provide services such as accommodation, transport services, tour guiding, eateries and restaurants, entertainment, and souvenir shops, which were not optimally exploited because locals were not trained and lacked the capacity to fully benefit from the activity.

It was also observed that there was a lack of capacity on part of the public bodies in terms of expertise, financial resources and organisational structure. As Tosun had pointed out in his work, administrative bodies in developing nations might not want to spend their limited resources on organising community participation whose benefits appear to be long drawn [15]. This is a major challenge faced in encouraging any community participation in decisions making especially for nations like India with a large and diverse population. The entire exercise of community participation from identifying stakeholders to actually including their suggestions in policies might take months of bureaucratic work as well as financial resources. This further discourages any innovation in the decision-making process.

5.4 Cultural limitations

Decision making is typically seen as the realm of the government and not people. Despite a democratic system of governance, community's understanding of democracy is limited to their voting rights. As most of the community members were also poor and socially marginalised, there was a fear that "interrupting" the system would be used against them in the future by the authorities. There was also a belief among people that authorities are best suited to make decisions and therefore common folk should not interfere with their working.

Despite an overall enthusiastic attitude towards tourism for its contribution towards economic development, there were also apprehensions about its negative impacts among locals. A stronger resistance to tourist influx into the areas was seen among older residents who feared that their traditional culture and lifestyle would be compromised. Participants also felt that authorities would give greater attention to the needs and wants of the tourists rather than the locals. There is already an increasing "museumification" of the areas where they are preserved to suit the sensibilities of the tourists rather than take into account the needs of the locals. For example, a resident complained that heritage homeowners in Shahjahanabad area cannot make any changes to their houses (such as installation of air conditioners). Due to such restrictions and no positive reinforcement from the authorities, the attitude of locals has changed against heritage and a growing number of residents wanted to get rid of their heritage houses. Such personal problems associated with the heritage nature of the area and thus policies created an animosity and thus further prevented people from enthusiastically and meaningfully participating in heritage and tourism management.

6 CONCLUSION

Existing barriers to community participation not only prevent residents from participating in decision making but also discourages them from engaging in any such exercise in the future. This then creates an uncomfortable relationship between local residents and tourism which ultimately leads to animosity towards heritage and conservation activities.

There is an urgent need to work towards meaningful integration of the communities into heritage tourism planning process. First of all, there needs to be an awareness drive to educate the masses about the importance of heritage, impact of tourism and their role in managing their resources. This can be followed by building local citizens networks and groups to engage residents, shopkeepers, traders, NGOs and other relevant stakeholder groups. Such groups would function to strengthen local democracy and allow citizens to voice their opinions. The authorities then shoulder the responsibility to legitimise such local area initiatives and invite them for discussions and deliberations on issues that might affect the locals. There is also a need for confidence building and communication between the governing agencies and local population to assure the latter that their concerns and suggestions would be taken into account. It is only when people start seeing responsive policies and development that they are likely to continue their engagement with policy making.

REFERENCES

[1] Chan, P.Y., Community participation in heritage management: A case study in Macau, New York. Master's thesis, Columbia University, 2016.

[2] World Travel and Tourism Council, *Travel & Tourism: City Travel and Tourism Impact 2018*, London, 2018.

[3] ICOMOS, *Heritage as a Driver for Development: Its Contribution to Sustainable Tourism in Contemporary Society*, Paris, 2011.

[4] Tosun, C., Expected nature of community participation in tourism development. *Tourism Management*, **27**(3), pp. 493–504, 2006.

[5] Huibin, X. & Marzuki, A., Community participation of cultural heritage tourism from innovation system perspective. *International Journal of Services Technology and Management*, **18**(3/4), pp. 105–127, 2012.

[6] Creaco, S., The role of tourism in sustainable economic development, *43rd Congress of the European Regional Science Association*, Jyväskylä, Finland, 2003.

[7] World Tourism Organisation and European Travel Commission, City Tourism and Culture – The European Experience, 2005. www.etccorporate.org/resources/uploads/ETC_CityTourism&Culture_LR.pdf. Accessed on: 26 Jan. 2020.

[8] UNESCO, 1972. https://whc.unesco.org/archive/convention-en.pdf. Accessed on: 5 Mar. 2020.

[9] Council of Europe, *Council of Europe Framework Convention on the Value of Cultural Heritage for Society*, Faro, 2005.

[10] Cole, S., Information and empowerment: The keys to achieving sustainable tourism. *Journal of Sustainable Tourism*, **14**(6), pp. 629–644, 2006.

[11] Snyman, S.L., The role of tourism employment in poverty reduction and community perceptions of conservation and tourism in southern Africa. *Journal of Sustainable Tourism*, **20**(3), pp. 395–416, 2012.

[12] Haywood, M., Responsible and responsive tourism planning in the community. *Tourism Management*, **9**(2), pp. 105–118, 1988.

[13] Krippendorf, J., Towards new tourism policies: The importance of environmental and sociocultural factors. *Tourism Management*, **3**(3), pp. 135–148, 1982.

[14] UNCED, Agenda 21, The Rio Declaration on Environment and Development, Rio de Janerio, 1992.

[15] Tosun, C., Limits to community participation in the tourism development process in developing countries. *Tourism Management*, **21**, pp. 613–633, 2000.

[16] Miles, M.B., Qualitative data as an attractive nuisance: The problem of analysis, *Administrative Science Quarterly*, **24**(4), pp. 590–601, 1979.

[17] Chaiklin, H., Doing case study research. *American Journal of Dance Therapy*, **22**, pp. 47–59, 2000.

[18] Flick, U., *An Introduction to Qualitative Research*, Sage Publications: London, 2009

[19] Glaser, B. & Strauss, A.L., *The Discovery of Grounded Theory: Strategies for Qualitative Research*, Aldine: New York, 1967

[20] Aga Khan Development Network, Nizamuddin Urban Renewal Initiative: Annual Report, New Delhi, 2015.

[21] The Economic Times, 2019. https://economictimes.indiatimes.com/news/politics-and-nation/tourists-up-at-taj-mahal-and-red-fort-but-qutub-minar-loses-its-no-2-spot/articleshow/70152555.cms?from=mdr. Accessed on: 20 Dec. 2019.

[22] Shahjahanabad Redevelopment Corporation, Revitalization of Shahjahanabad-Project Concept Proposal, New Delhi, 2014.

[23] Hindustan Times, 2018. www.hindustantimes.com/delhi-news/a-workshop-for-heritage-structure-owners-of-shahjahanabad-on-how-to-renovate-properties/story-7DzqpH278bsyfNSxT0NdbO.html. Accessed on: 2 Jan. 2020.

[24] Kala, D. & Bagri, S.C., Barriers to local community participation in tourism development: Evidence from mountainous state Uttarakhand, India. *Tourism*, **66**(3), pp. 318–333, 2018.

Author index

City Sustainability and Regeneration

Edited by: ***S. MAMBRETTI****, Polytechnic of Milan, Italy and* ***J. L. MIRALLES I GARCIA,*** *Politechnic University of Valencia, Spain*

A set of new studies are included in this volume which provides solutions that lead towards sustainability. Contributions originate from a diverse range of researchers, resulting in a variety of topics and experiences.

Urban areas face a number of challenges related to reducing pollution, improving main transportation and infrastructure systems and these challenges can contribute to the development of social and economic imbalances and require the development of new solutions. The challenge is to manage human activities, pursuing welfare and prosperity in the urban environment, whilst considering the relationships between the parts and their connections with the living world. The dynamics of its networks (flows of energy matter, people, goods, information and other resources) are fundamental for an understanding of the evolving nature of today's cities.

Large cities represent a productive ground for architects, engineers, city planners, social and political scientists able to conceive new ideas and time them according to technological advances and human requirements. The multidisciplinary components of urban planning, the challenges presented by the increasing size of cities, the amount of resources required and the complexity of modern society are all addressed.

ISBN: 978-1-78466-415-2 eISBN: 978-1-78466-416-9
Published 2020 / 164pp

www.ingramcontent.com/pod-product-compliance
Lightning Source LLC
LaVergne TN
LVHW062232200926
848484LV00006B/1523